International Conference on Smart Materials Applications (ICSMA 2018)

AA006792

IOP Conference Series: Materials Science and Engineering Volume 422

Singapore
26 – 28 January 2018

ISBN: 978-1-5108-7506-7
ISSN: 1757-8981

Printed from e-media with permission by:

Curran Associates, Inc.
57 Morehouse Lane
Red Hook, NY 12571

Some format issues inherent in the e-media version may also appear in this print version.

Printed by Curran Associates, Inc. (2019)

For permission requests, please contact the Institute of Physics
at the address below.

Institute of Physics
Dirac House, Temple Back
Bristol BS1 6BE UK

Phone: 44 1 17 929 7481
Fax: 44 1 17 920 0979

techtracking@iop.org

Additional copies of this publication are available from:

Curran Associates, Inc.
57 Morehouse Lane
Red Hook, NY 12571 USA
Phone: 845-758-0400
Fax: 845-758-2633
Email: curran@proceedings.com
Web: www.proceedings.com

TABLE OF CONTENTS

IOP Conf. Series: Materials Science and Engineering **422** (2018) 011001 doi:10.1088/1757-899X/422/1/011001

Preface

2018 International Conference on Smart Materials Applications (ICSMA 2018) was held at Shaw Foundation Alumni House of National University of Singapore during 26-28 January 2018. The purpose of this conference was to provide a forum for accessing to the most up-to-date and authoritative knowledge from both industrial and academic worlds, sharing best practice in the field of smart materials analysis. The meeting provided an opportunity to highlight recent developments and to identify emerging and future areas of growth in this exciting field.

As a platform for technical exchange, each author has 15 minutes to introduce his or her research. In addition to the presentation, each author is encouraged to connect and exchange ideas with other scholars. Each contributed paper has been peer-reviewed by reviewers who were collected organizing and technical committee members as well as other experts in the field from different countries. The proceedings is not intended to be exhaustive about all areas of smart material applications. Rather, it focuses on some selected topics, especially material physics and computational materials science.

There are various topics including engineering polymers, functional materials, nuclear fuel materials, biomaterials, sensors and surfaces, magnetic properties, corrosion, erosion, wear resistance, electron microscopy, x-ray phase analysis. After several rounds of the review procedure, the program committee accepted papers for the publication in materials science forum.

We would like to thank the organizing committee and the members of reviewers for their kind assistances in reviewing the papers. We would also extend our best gratitude to Prof. LU Li, National University of Singapore, Singapore; Prof. Hao Gong, National University of Singapore, Singapore; Prof. Alfonso Maffezzoli, University of Salento, Italy; and Assoc. Prof. GUPTA Manoj, National University of Singapore, Singapore for their invaluable contribution and worthwhile ideas shared in the ICSMA 2018 conference.

Much of the credit of the success of the conference is due to topic coordinators who have devoted their expertise and experience in promoting and in general co-ordination of the activities for the organization and operation of the conference. The coordinators of various session topics have devoted a considerable time and energy in soliciting papers from relevant researchers for presentation at the conference.

We express our deepest gratitude to all members of the scientific committee, reviewers, sponsors, participants in the conference and members of the ICSMA organizing committee for their valuable contribution to the successful organization of ICSMA 2018. For further information about ICSMA, please visit the website www.icsma.org.

Conference Committee Chair
Prof. Xiaohong Zhu,
Sichuan University, China
January 28, 2018

IOP Conf. Series: Materials Science and Engineering **422** (2018) 011002 doi:10.1088/1757-899X/422/1/011002

CONFERENCE COMMITTEES

Conference Chairs
Prof. LU Li, National University of Singapore, Singapore
Prof. Xiaohong Zhu, Sichuan University, China

Program Committee Chairs
Prof. Alfonso Maffezzoli, University of Salento,Italy
Assoc. Prof. Gupta Manoj, National University of Singapore, Singapore
Dr. Xuelin Lei, East China University of Science and Technology, China

Technical Committee
Assoc. Prof. Mohammad Sukri Mustapa, Universiti Tun Hussein Onn Malaysia (UTHM), Malaysia
Assoc. Prof. Chou Chau Yuan-Fong, Universiti Brunei Darussalam,Brunei Darussalam
Dr. Mohd Rasidi Bin Haji Ibrahim, Universiti Tun Hussein Onn Malaysia, Malaysia
Karel Raz, University of West Bohemia, Czech Republic
Ahmad Mubarak Tajul Arifin, Universiti Tun Hussein Onn Malaysia (UTHM), Malaysia
Dr. Koorosh Gharehbaghi, RMIT University, Australia
Prof. Dr. Rupinder Singh, Guru Nanak Dev Engineering College, India
Prof. Lung-Ming Fu, National Pingtung University of Science and Technology, Taiwan
Prof. Dr. Masayuki Yamaguchi, Japan Advanced Institute of Science and Technology, Japan
Dr. Alicja Klimkowicz, Shibaura Institute of Technology, Japan
Prof. Mohammad Wani, NIT Srinagar, India
Prof. Pramuan Tangboriboonrat, Mahidol University, Thailand
Asst. Prof. Dr. Teeraporn Suteewong, King Mongkut's Institute of Technology Ladkrabang, Thailand
Dr.Tomáš Homola, Masaryk University, Czech Republic
Asst. Prof. Chih-Ling Huang, Kaohsiung Medical University, Taiwan
Assoc. Prof. Teik-Cheng Lim, Singapore University of Social Sciences, Singapore
Dr. Huynh Trong-Phuoc, Cantho University (CTU), Vietnam
Assoc. Prof. Achanai Buasri, Chulalongkorn University, Thailand
Asst. Prof. Dr. Francesca Scalisi, University of Palermo, Italy
Prof. Xiaolin Wang, University of Wollongong, Australia
Prof. Chong Su Cho, Seoul National University, South Korea
Prof. Sivakumar Manickam, The University of Nottingham Malaysia Campus, Malaysia
Prof. Dr. Yu-Chieh Cheng, National Taipei University of Technology (NTUT), Taiwan

IOP Conf. Series: Materials Science and Engineering **422** (2018) 011003 doi:10.1088/1757-899X/422/1/011003

Peer review statement

All papers published in this volume of *IOP Conference Series: Materials Science and Engineering* have been peer reviewed through processes administered by the proceedings Editors. Reviews were conducted by expert referees to the professional and scientific standards expected of a proceedings journal published by IOP Publishing.

IOP Conf. Series: Materials Science and Engineering **422** (2018) 012001 doi:10.1088/1757-899X/422/1/012001

Finite-Element Modeling of Particle Size Effect on Mechanical Properties of SiCp/Fe Composites

J Y Cao[1], J F Jin[1,*], L Wang[1], T Cheng[1] and P J Yang[1]

[1] School of Material Science and Engineering & Key Laboratory for Anisotropy and Texture of Materials, Ministry of Education, Northeastern University, Shenyang, 110819, China

E-mail: neu_caojy@163.com, jinjf@atm.neu.edu.cn, wangl_edu@163.com chta0929@qq.com, yangpj_yangcx@163.com

Abstract: Particle size has a significant effect on mechanical properties of particle reinforced metal matrix composites (PRMMCs). Here, the effect of particle size on mechanical properties of SiCp/Fe composite has been studied using a finite element (FE) model incorporated with the Taylor-based nonlocal theory (TNT) of plasticity, where the tested particle size is 5, 10, 16, 21, 28 and 40μm, respectively. The results indicate that the TNT-FE model overcomes the shortage of the traditional FE model, which cannot deal with the intrinsic size effect. With the particles of 20 vol.%, the simulated flow stress of the iron matrix composite reinforced by 16μm SiC particles is the highest and then follows the order of 28 > 5 > 10 > 21> 40μm particle reinforced ones, close to the tendency of experimental results. A large area of compressive stress zone is observed in the matrix near the particles in the composites reinforced by 16 and 28μm particles, which implies that the matrix can be well protected during loading, so that the load-bearing capability is better. The purpose is to optimize mechanical property of SiCp/Fe composite by adjusting particle size and eventually to design PRMMCs from microstructural control.

1. Introduction

Particle reinforced metal matrix composites (PRMMCs) have high modulus, strength, high wear resistance and thermal stability [1, 2]. It is observed that mechanical properties of the PRMMCs depend not only on the type of reinforced particles, but also on the size, content, shape and distribution of particles and the features of the particles/matrix interface [3-5]. In Zhuang's work [6], mechanical properties of the iron matrix composite reinforced by 13μm SiC particle was the best with 10 vol.% SiC. Similarly, Zhang *et al.* observed a peak strength of the 13μm SiCp/Fe composite when the particle size increased from 3.5μm to 45μm with 20 vol.% copper-coated SiC reinforcements [7]. Therefore, to understand the relationship between strengthening behavior and microstructure is a critical issue in developing PRMMCs. The continuum models have been used to study mechanical properties of PRMMCs, such as finite element (FE) model [8-10]. However, traditional FE models are based on classical plasticity theory, which lead to a dependence of deformation behavior on volume fraction of reinforcing particles but not on particle size [11]. Recently, dislocation models and Taylor-based nonlocal theory (TNT) of plasticity continuum models have been developed to understand the particle size effect [12-14]. Here, the effect of particle size on mechanical properties of SiCp/Fe composite is studied using a TNT FE model, where the tested particle size is 5, 10, 16, 21, 28 and 40μm, respectively, with 20 vol.% particles.

2. Finite Element Model Setup and Taylor-based nonlocal theory (TNT) of plasticity

In this work, the plane-strain 2D finite element model is employed to simulate a multi-particle composite system during monotonic tensile loading and the model setup for SiCp/Fe composite shows in Figure.1 (a), with spherical particles embedded into the rectangular matrix (ABCD). The volume fraction of SiC particles is 20% and the tested particle size is 5, 10, 16, 21, 28 and 40μm, respectively, corresponding to the particle number from 64 to 1 and particles distributed uniformly, shown in Figure.1 (b).

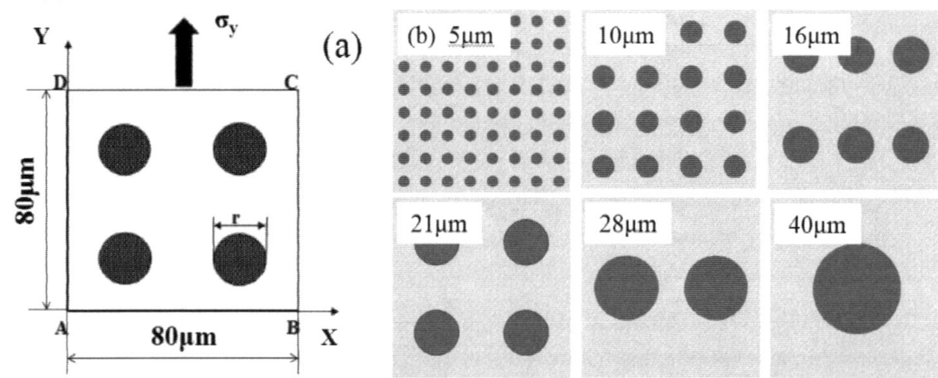

Figure 1. (a) Schematic of the plane-strain 2D finite element model for SiCp/Fe composite and (b) the model setup with different size particles, the tested particle size is from 5 to 40μm, corresponding to the particle number from 64 to 1 and particles distributed uniformly (Pink: SiC particle and blue: iron matrix) .

Based on the Taylor nonlocal theory (TNT) of plasticity, the shear flow stress τ can be given by:

$$\tau = \alpha G_{m} b(\rho)^{1/2} \tag{1}$$

where G_m is shear modulus, b is Burgers vector, α is an empirical coefficient and ρ is dislocation density. In this work, the ρ is only composed of geometrically necessary dislocations (GNDs) due to geometrical mismatch (GM) generated in the composite matrix as a result of deformation-induced plastic strain gradient [15, 16]. Some researchers induced particle size effect into various continuum models by incorporating dislocation plasticity to alter the flow stress in the composite matrix (σ_{Matrix}^{Flow}), which achieved good results [13, 14, 17]. Therefore, the $\sigma_{\text{Matrix}}^{\text{Flow}}$ can be given as

$$\sigma_{\text{Matrix}}^{\text{Flow}} = \sigma_{\text{Matrix}}^{\text{Ref}}(\varepsilon_{\text{p}}) + M\alpha G_{\text{m}} b(\rho_{\text{GNDs}}^{\text{GM}})^{1/2} \tag{2}$$

where $\sigma_{\text{Matrix}}^{\text{Ref}}(\varepsilon_{\text{p}})$ is the stress in unreinforced matrix under plastic strain of ε_{p}, M is the Taylor factor, acting as an isotropic interpretation of the crystalline anisotropy at the continuum level and $\rho_{\text{GNDs}}^{\text{GM}}$ is the GNDs due to the GM. An enhanced FEM model incorporating the TNT of plasticity has been developed and successfully used in studying the particle size dependent flow strengthening in PRMMCs[14], in which assuming that the particle was pure elastic and the matrix was elastoplastic, the flow stress of the reinforced matrix in Eq.(2) could be adapted as follows,

$$\sigma_{\text{Matrix}}^{\text{Flow}} = \sigma_{\text{Matrix}}^{\text{Ref}}(\varepsilon_{\text{p}}) + [27\times(5/2)^{1/2}\alpha^2 G_{\text{m}}^2 \frac{f^{1/3}}{r} b\varepsilon_{\text{p}}]^{1/2} \tag{3}$$

where f and r was volume fraction and size of particle, respectively.

In this work, the $\sigma_{\text{Matrix}}^{\text{Flow}}$ in Eq.(3) is used as the numerical constitutive input of elastoplastic iron matrix into the TNT-FE model, which considers the effects of intrinsic size and volume fraction of particles on the matrix properties. The SiC reinforcement is treated as pure elastic particle in the FE model. The model setup for the SiCp/Fe composite has already been introduced earlier in Figure.1, and the interface between the Fe and SiC assumes to be connected perfectly. The ANSYS commercial software has been used to implement the FE simulations.

3. Results and Discussion

In this section, particle size effect on mechanical behaviours of the SiCp/Fe composites was studied using the plane-strain 2D finite element (FE) model incorporating with Taylor-based nonlocal theory (TNT) of plasticity.

Firstly, the parameter $\boldsymbol{\alpha}$ in Eq.(3) for the SiCp/Fe composite is determined to 0.4 from experimental data fitting of tensile strength in 21μm SiCp/Fe with 20 vol.% copper-coated particles [18] and the simulated tensile strength is 535MPa, which is only 3MPa lower than the experimental one of 538MPa, shown in Figure.2 (a). Moreover, the tensile strength of 10μm SiCp/Fe has also been tested with 562MPa from our model, compared with the experimental value of 567MPa of 13μm SiCp/Fe. Therefore, this model is valid to use for predicting other-sized SiC reinforced iron matrix composites with $\boldsymbol{G_m}$ of 67GPa and $\boldsymbol{b}$ of 2.5×10^{-10}m of pure iron in Eq.(3).

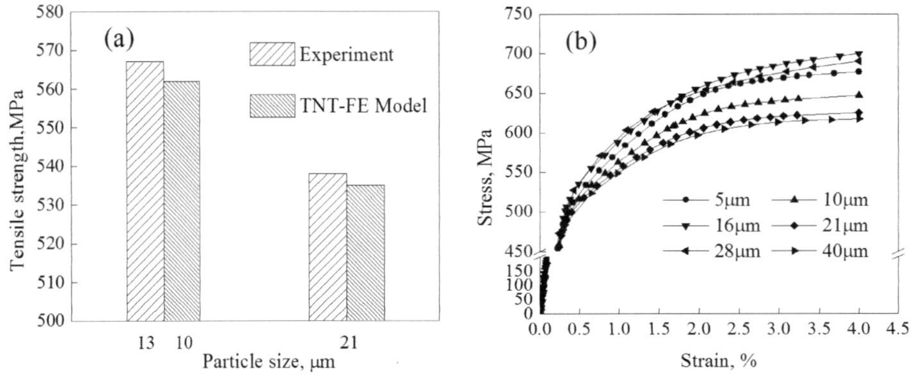

Figure 2. (a) The comparison of experimental and simulated values of tensile strength in 13μm (experiment) via 10μm (simulated) and 21μm SiCp/Fe composites and (b) the simulated stress-strain curves of SiCp/Fe with different particle size.

In Figure.2 (b), the simulated stress-strain curves of the SiCp/Fe composites with different particle size are shown, which depict a size-dependent strengthening phenomenon. The simulated flow stress at the applied strain of 4% from traditional FE and TNT FE models are compared in Figure.3 (a). The simulated stress from traditional model is almost the same in the size of 5, 10, 21 and 40μm cases, and in the size 16 and 28μm cases, respectively, which means that the traditional FE model cannot deal with the intrinsic size effect.

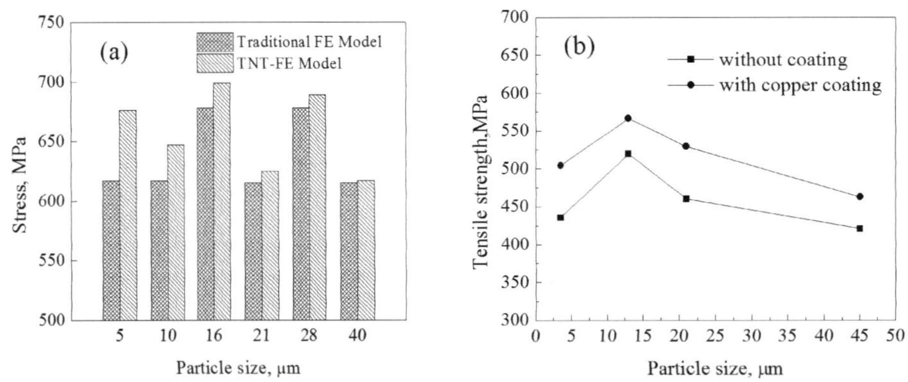

Figure 3. (a) Flow stress at the applied strain of 4% versus the particle size from traditional and TNT FE models and (b) tensile strength versus particle size of SiCp/Fe from experimental results. [7]

In Figure.3 (a), the simulated flow stress at the applied strain of 4% from the TNT FE model is 676, 647, 699, 625, 689 and 617MPa, respectively, associated with particle size of 5, 10, 16, 21, 28 and 40μm. The stress of the composite reinforced by 16μm particles is the highest and then follows the order of 28 > 5 > 10 > 21 > 40μm particles reinforced ones, close to the tendency of tensile strength from experimental observations [7] in Figure.3 (b), where the tested particle was 3.5, 13, 21 and 45μm, respectively.

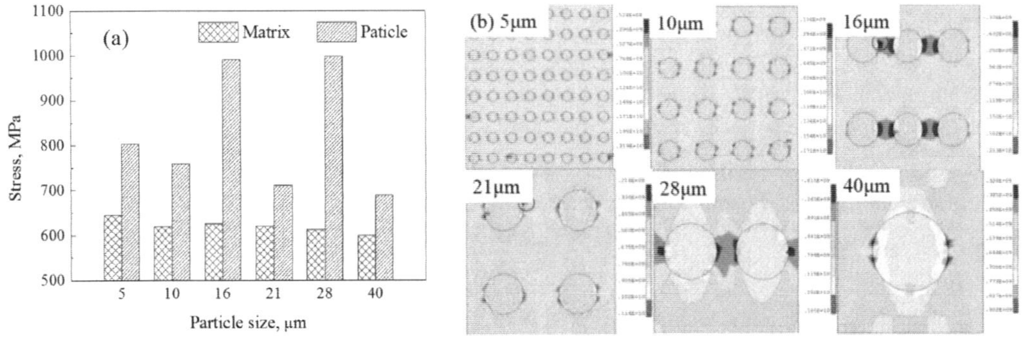

Figure 4. At the applied strain of 4%, (a) the average normal stress in the particles and matrix in loading direction (y-axis) and (b) the normal stress (σ_{yy}) contour of the SiCp/Fe composites with different particle sizes.

In Figure.4 (a), the average normal stress in the particles and matrix in the loading direction at the applied strain of 4% has been compared, associated with different particle size. The stress in the matrix decreases with the increasing particle size, which means that the strengthening on the matrix from the particles is better when the particle size is smaller. The stress in the particles for the composite reinforced by 16 and 28μm particles is greater, which means that the load-bearing capability of such particles is better. Furthermore, The normal stress (σ_{yy}) contours of the SiCp/Fe composites with different particle sizes at 4% strain are plotted in Figure.4 (b). A large area of compressive stress zone exists in the matrix near the particles in the composites reinforced by 16 and 28μm particles, which implies that the matrix can be well protected during loading so that the load-bearing capability is better.

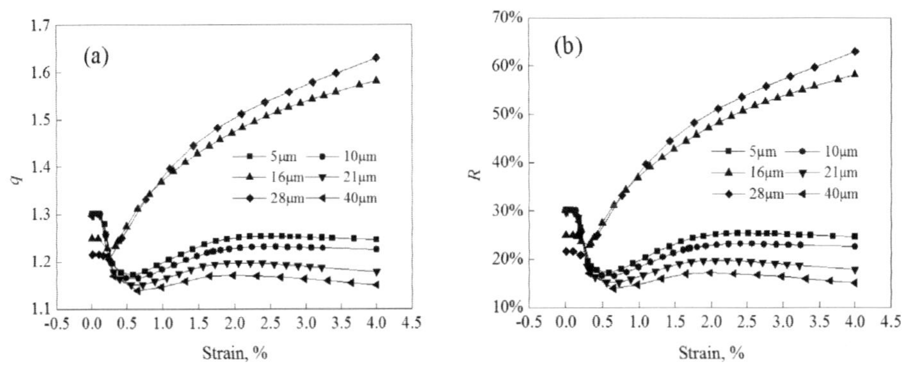

Figure 5. The particle size effect (a) on phase stress partition parameter (q) and (b) on strengthening efficiency (R) during straining.

The stress relationship between the matrix and the particle can be described by the phase stress partition parameter (q) given by the equation of $q = \sigma_p/\sigma_m$ [12], where σ_m is stress in matrix and σ_p is stress in particles. The relationship between the q and applied strain is plotted in Figure.5 (a). It is shown that for the 16 and 28μm SiCp/Fe composites, the q decreases slightly at first after elastic deformation, then increases dramatically and is over 1.60 at 4% strain, which is greater than the value of 1.40 in 1~55μm SiCp/Al composites with 20 vol.% SiC [12]. While, the q decreases sharply and then becomes relatively stable in 1.20 for the 5, 10, 20 and 40μm SiCp/Fe composites. Moreover, the strengthening efficiency (R) of a given volume percentage of reinforcement on the matrix can be performed by the equation of $R = (\sigma_c - \sigma_m)/V\sigma_m$[19], where V is volume fraction of particles, σ_m is stress in matrix and σ_c is the composite stress. The relationship between the R and applied strain is plotted in Figure.5 (b), where the R can reach about 60% for the 16 and 28 μm SiCp/Fe composites, while it keeps about 20% for the others, comparing with the R of 2.4% in SiCp/Al, 20.3% in carbon nanotube reinforced Cu and 94% in graphene nanosheets reinforced Cu composites [19].

4. Conclusions

Using a plane-strain 2D finite element (FE) model by incorporating Taylor-based nonlocal theory (TNT) of plasticity, the effect of particle size on mechanical properties of SiCp/Fe composite is studied, where the tested particle size is 5, 10, 16, 21, 28 and 40μm, respectively, with 20 vol.% particles.

1. By comparison with experimental results, the TNT FE model overcomes the shortage of the traditional FE model which cannot deal with the intrinsic size effect.
2. The simulated flow stress of the iron matrix composite reinforced by 16μm SiC particles is the highest and then follows the order of 28 > 5 > 10 > 21 > 40μm particle reinforced ones, close to the tendency of experimental observation.
3. A large area of compressive stress zone is observed in the matrix near the particles in the composites reinforced by 16 and 28μm particles, which implies that the matrix can be well protected during loading, so that the load-bearing capability is better.

5. Acknowledgements

Authors would like to acknowledge the National Key Research and Development Program of China (No.2016YFB0701204), Projects by National Natural Science Foundation of China (No.U1302272), National Higher-education Institution General Research and Development Project (No.N151004004, N150502002) and Program for Liaoning Excellent Talents in University (No. LJQ2015037).

6. References

[1] Chawla N and Chawla K K. 2005 *Metal Matrix Composites* (New York: Springer)
[2] Zhang G S, Wei S Z, Han M R, et al. 2013 *Particle Reinforced Steel Matrix Composites* (Bei Jing: Science Press).
[3] Wang Y M, Zong Y P, Zhuang W B, et al. 2011 *Materials Science and Technology* **19** 47-51
[4] Li J, Zong Y P, Wang Y M, et al. 2011 *Journal of Materials Science and Engineering* **29** 321-6
[5] Zhang Y B, Zong B Y, Jin J F, et al. 2015 *Surf. Eng.* **31** 232-9
[6] Zhuang W B, Zong B Y, Wang Y M, et al. 2013 *J. Compos Mater.* **47** 1001-10
[7] Zhang Y B, Zong B Y, Jin J F, et al. 2014 *Applied Mechanics and Materials* **556-562** 302-5
[8] Shen H and Lissenden C J 2002 *Mater. Sci. Eng., A.* **338** 271-81
[9] Chawla N, Ganesh V and Wunsch B 2004 *Scr. Mater.* **51** 161-5
[10] Mishnaevsky L L 2004 *Acta Mater.* **52** 4177-88
[11] Shao J Y and Liu Y 2007 *Materials Review* **21** 111-5
[12] Yan Y W, Geng L and Li A B 2007 *Mater. Sci. Eng., A.* **448** 315-25
[13] Dai L, Ling Z and Bai Y 2001 *Compos. Sci. Technol.* **61** 1057-63

[14] Shao J C, Xiao B L, Wang Q Z, et al. 2011 *Compos. Sci. Technol.* **71** 39-45

[15] Gao H, Huang Y, Nix W D, et al. 1999 *J. Mech. Phys. Solids.* **47** 1239-63

[16] Gao H and Huang Y 2001 *Int. J. Solids Struct.* **38** 2615-37

[17] Wang Y M, Zhang C H, Zong Y P, et al. 2013 Adv. *Compos. Mater.* **22** 299-310

[18] Zhang Y B 2015 *Effects of Metal Coating on SiC Reinforcing Particles and High-Energy Ball Milling on Properties of SiCp/Fe Composites*, PhD thesis (Shen Yang: Northeastern University)

[19] Tang Y, Yang X, Wang R, et al. 2014 *Mater. Sci. Eng., A.* **599** 247-54

IOP Conf. Series: Materials Science and Engineering **422** (2018) 012002 doi:10.1088/1757-899X/422/1/012002

Research on Pin-load Distribution for Countersunk Joint Considering Accumulated Damage of Composite Laminates

X D Liu[1], H M Zhang[1], N Liu[2] and L J Li[1]

[1] Department of Mechanics, Xi'an University of Science and Technology, Xi'an, China
[2] Department of Building Environment and Energy Application Engineering, Xi'an University of Science and Technology, Xi'an, China

E-mail: 18192194331@163.com, 1522793926@qq.com

Abstract. The paper presents a numerical simulation about the damage process of the multiple countersunk bolted double-lap joints based on the cumulative damage theory which is widely used to estimate the mechanical strength of fastener joints, the Hashin's failure criterion, and the Tan's stiffness degradation model, and investigates the influence of the composite materials cumulative damage on the pin-load distribution. The numerical model of composite joints has been established in the commercial finite element software ANSYS to research the influence of the composite cumulative damage on the pin-load distribution, to assess the feasibility of integral analysis as well as to evaluate the error and rationality of traditional methods by applying the global analysis technology. The conclusions demonstrate that the results obtained with the method of composite cumulative damage considering properties degradation are close to those of conventional methods. The pin-load distribution is characterized by "Bolts at both ends bearing greater load, the central bolts bearing smaller load" during the whole loading process. Using conventional methods to study the pin-load distribution is easy and efficient to meet the needs of engineering application.

1. Introduction

Composite materials have been widely applied in the aircraft load-bearing structure owing to their excellent mechanical properties. The connection types between composite material components and other components are gluing, mechanical connection, and mixed connections primarily. The most common connection is a mechanical joint which possesses the characteristics of simple process, reliable connection, easy disassembly and maintenance, and transmitting heavy load. Unlike glued structures, in structures with bolted joints it is required to drill holes in the composite. The openings destroy the continuity of the fiber, result in stress concentration near the holes, and further impair the load-bearing capacity of the composite structures. The connection strength of composite materials directly affects the safety and service life of aviation structures. The stress concentration around holes is more serious than that of the metal structures due to the anisotropy of the composite material. Therefore, determining pin load is the basis for calculating the strength of the structures with the common multiple-bolted joints.

The methods of determining the pin-load proportion of composite materials with multiple-bolted joints mainly include the test method, analytical method and finite element method, respectively [1-4]. The brittle nature of the composite is not conducive to pin-load redistribution and homogenization, thus, for the static analysis of the mechanical connection of the composite structures, the present research usually does not consider the effect on pin-load distribution due to material damage leading to its properties degradation under the external load, and only considers the initial state of the pin-load distribution, regardless of its changes with the applied load; In the ideal condition, the pin-load distribution remains essentially constant[1-8], however, this idealized result is obtained by an approximately linear analysis that does not consider the composite damage process. In fact, as the external load increases, the gradual accumulation of bearing and other damage will lead to changes of bearing stiffness around in pin-hole zone and affect the overall load distribution.

A series of experimental results show that the pin-load distribution will change with the change of the external load after the occurrence of pin-hole bearing damage. Therefore, it is necessary to consider the effect of the composite material on pin load after the damage. This paper intends to use three-dimensional cumulative damage model analysis technology, start from the overall analysis to study the effect of composite material cumulative damage on the pin-load distribution, reveal the feasibility of the overall analysis, and estimate the error and reasonability of traditional methods.

2. Cumulative damage analysis method

When exerting the numerical method to simulate the process of damage accumulation, it is needed to determine whether the material is damaged or not at the first. Secondly, the variables that quantitatively describe the damage are introduced and correlated with the macroscopic mechanical behavior of the material, that is, the damage constitutive relationship is to be established and material degradation will be implemented. Finally, the failure criteria of structures are proposed. In summary, material failure criterion, damage constitutive relation, and destruction criterion are called three elements of the cumulative damage criterion in the process of the cumulative damage simulation in composite structures. Therefore, it is important to choose the reasonable failure criterion and damage constitutive relation for the numerical simulation. In the analysis of composite structures, there are more than ten kinds of theories about composite strength [9-12], among them, the most commonly used criteria include Hashin's criterion, Tsai-Hill criterion and the maximum stress criterion and others.

2.1 Failure criteria

In the analysis of composite material damage, the failure criterion of the stress quadratic polynomial for the composite lamina was proposed by Hashin[12]. Hashin's criterion is also widely used to the analysis of structural damage of composite material, and has been developed continuously in the applications, and a variety of Hashin's criteria derivatives are developed. Through the study of composite material structure strength analysis, it is found that delamination damage usually appears firstly in damage mode, and reduces the compression strength of composite laminated plates and the impact resistance strength during the loading process of composite laminated plate. Hashin's failure criteria considering the delamination damage was shown in Table 1[13].

Table 1. Three dimensional Hashin's failure criteria for composite materials based on delamination

Failure Mode		Hashin's Failure Criteria
Fiber tensile cracking	$\sigma_{11} \geq 0$	$(\sigma_{11}/X_T)^2 + (\sigma_{12}/S_{12})^2 + (\sigma_{13}/S_{13})^2 \geq 1$
Fiber compressive cracking	$\sigma_{11} < 0$	$(\sigma_{11}/X_C)^2 \geq 1$
Matrix tensile cracking	$\sigma_{22} \geq 0$	$(\sigma_{22}/Y_T)^2 + (\sigma_{12}/S_{12})^2 + (\sigma_{23}/S_{23})^2 \geq 1$
Matrix compressive cracking	$\sigma_{22} < 0$	$(\sigma_{22}/Y_C)^2 + (\sigma_{12}/S_{12})^2 + (\sigma_{23}/S_{23})^2 \geq 1$
Fiber-matrix shear out	$\sigma_{11} < 0$	$(\sigma_{11}/X_C)^2 + (\sigma_{12}/S_{12})^2 + (\sigma_{13}/S_{13})^2 \geq 1$
Delamination in tension	$\sigma_{33} > 0$	$(\sigma_{33}/Z_T)^2 + (\sigma_{13}/S_{13})^2 + (\sigma_{23}/S_{23})^2 \geq 1$

Delamination in compression	$\sigma_{33}<0$	$(\sigma_{33}/Z_C)^2+(\sigma_{13}/S_{13})^2+(\sigma_{23}/S_{23})^2\geq 1$

2.2 Material properties degradation after damage

With the increase of load, the fiber and matrix of laminates will be gradually destroyed and the rigidity of the structural will be decreased. The degradation of composite elastic constants can be used to characterize the development of material damage. The commonly used method of material properties degradation assumes that if one of the elements is deteriorating, the degradation of material properties affects only the element itself without affecting other elements. The proper choice of parameter degeneration method has great influence on the ultimate strength of the laminated plate to be solved.

The parameter degradation method used by Chang[14] is that the material constants are degenerated to zero whenever failure occurs. Taking into account the convergence of the finite element calculation, these elastic constants are set to a very small value during the calculation. Tan [15] proposed material stiffness decrease caused by different damage modes is expressed with damage state variables in the planar case and different elastic constants degraded in proportion, which is extended to three-dimensional situation by Camanho and Matthews[1]. The two kinds of degradation rules are shown in Table 2.

Table 2. Material property degradation rules

Failure Mode	Chang's rule	Tan's rule
Matrix tensile cracking	$E_{22}=\nu_{12}=0$	$E_{22}\rightarrow 0.2E_{22}$, $G_{12}\rightarrow 0.2G_{12}$, $G_{23}\rightarrow 0.2G_{23}$
Matrix compressive cracking	$E_{22}=\nu_{12}=0$	$E_{22}\rightarrow 0.4E_{22}$, $G_{12}\rightarrow 0.4G_{12}$, $G_{23}\rightarrow 0.4G_{23}$
Fibre tensile failure	$E_{11}=E_{22}=E_{33}=G_{12}=G_{23}=G_{13}=\nu_{12}=\nu_{23}=\nu_{13}=0$	$E_{11}\rightarrow 0.07E_{11}$
Fibre compressive failure	$E_{11}=E_{22}=E_{33}=G_{12}=G_{23}=G_{13}=\nu_{12}=\nu_{23}=\nu_{13}=0$	$E_{11}\rightarrow 0.14E_{11}$
Fibre-matrix shear-out	$G_{12}=\nu_{12}=0$	$G_{12}=\nu_{12}=0$
Delamination in tension	$E_{33}=G_{23}=G_{13}=\nu_{23}=\nu_{13}=0$	$E_{33}=G_{23}=G_{13}=\nu_{23}=\nu_{13}=0$
Delamination in compression	$E_{33}=G_{23}=G_{13}=\nu_{23}=\nu_{13}=0$	$E_{33}=G_{23}=G_{13}=\nu_{23}=\nu_{13}=0$

2.3 The process of cumulative damage analysis

The analysis procedure of cumulative damage prediction was as follows: first of all, the stress of laminates was analyzed, and then failure criteria was used to determine whether laminate element failed. If element failure occurs, its material properties are correspondingly degraded. Then, continued to apply the load and repeat the previous process until the overall structure failed.

Liu [8] studied the strength of composite laminate joints by experiment and numerical method. The calculated results manifested that when the load is small, the degradation results by Chang's or Tan's rules are both smaller, and the latter's is closer to the experimental results. Both Chang's and Tan's degradation methods are based on the tensile and compressive loading of opening plates. Tserpes[7] developed the three-dimensional finite element model to investigate the cumulative damage process of single-bolted composite and proved that Tan's material degradation rule was more reasonable than that of Chang's.

3 Analysis of cumulative damage influence on pin-load distribution

By using large-scale commercial finite element software ANSYS, the three-dimensional finite element model was built to calculate the pin-load distribution of multiple countersunk bolted double-lap joints.

3.1 Analysis example

The analysed specimen was the same as presented in [6], composite plates were manufactured from carbon fiber/epoxy material system (HTA7/6376[4]) with quasi-isotropic layups $[(\pm 45/0/90)_3]_s$ and $[(\pm 45/0/90)_6]_s$ for the outer and middle plates, respectively. The nominal ply thickness was 0.13 mm. The thickness of the composite laminates was 3.12 and 6.24 mm for the outer and middle plates, respectively. The composite plates were joined by six fasteners of titanium[4] Torque-set. The nominal diameter was 6 mm for all fastener systems. The specimen geometry and configurations of six fastener systems are shown in Figure 1.

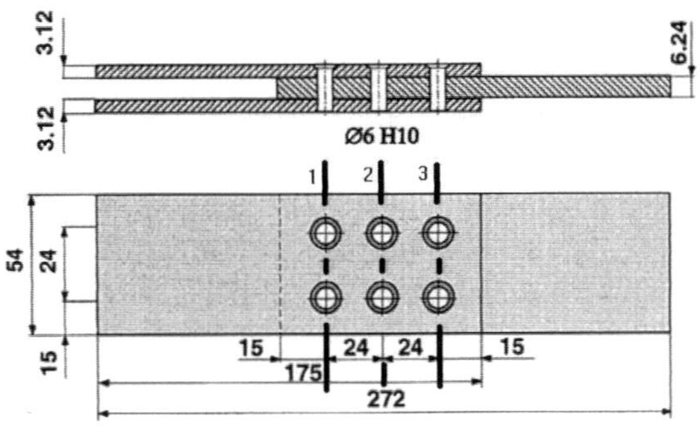

Figure 1. Dimension of test specimens with three row-bolt double-lap joint

3.2 The three-dimensional model for pin-load analysis

As shown in figure 2 and figure 3, bolts, nuts and gaskets were simplified as a part in this 3-D model which were meshed by 8-node solid element SOLID185, composite laminates were meshed by 8-node solid layer element SOLID185. A geometric model was meshed with a mapping method. In the model, the influence of contact friction and pretension force was also considered. The contact element between pin-hole, pin-plate and plate-plate adopted CONTA173 and the target element was TARGET170. The magnitude of the friction between the contact surfaces was controlled by Coulomb friction, and the friction coefficient was 0.2. The pretension force was applied by defining the prestress element PRETS179.

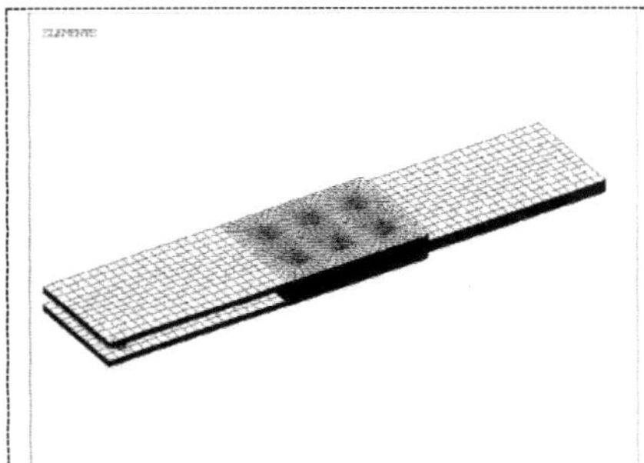

Figure 2. 3-D finite element mesh models

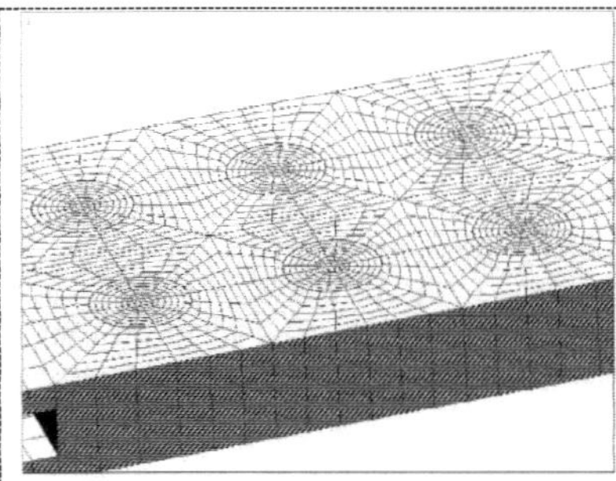

Figure 3. The modeling details of bolt-hole local mesh

The load was applied on the modeling in the same way as in the test, one end of the model (6.24 mm thick) was fixed and the other end was applied with displacement.

4 The results and discussion of pin-load distribution

4.1 Calculation Results

The resultant force of bearing pressure along the direction of the loading in pin-hole can be obtained by finite element method. The pretension load applied to the three-dimensional model was consistent with the experiment. The results of pin-load distribution of six countersunk bolts double-lap joints are shown in Figure 4.

As can be seen from Figure 4(a):

(1) At the beginning of loading, the change of 1-row, 2-row pin load was more obvious, the 1-row pin load increased gradually, and 2-row pin load decreased gradually under the condition of small load. When the load increased to a certain value, the pin load of each row tended to be stable.

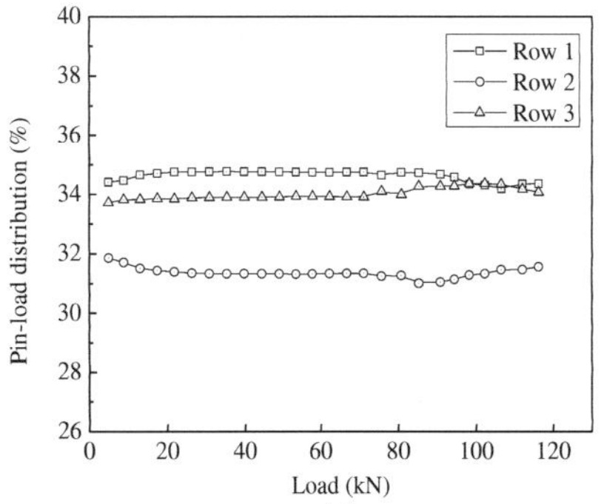

(a)

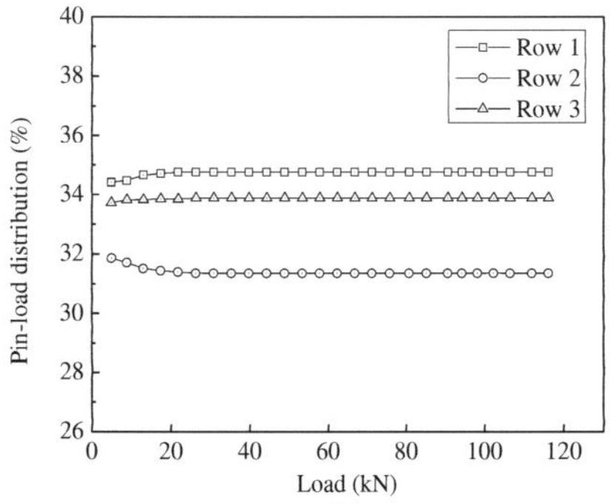

(b)

Figure 4. Pin-load distribution on double-lap of multiple-bolted joints: (a) with cumulative damage; (b) without cumulative damage

This was that the frictional force at the contact surface of the structure was much greater than the load on the pin in the early stage of loading. The external load was mainly used to overcome the friction, and the change of load distribution was obvious. With the further increase of the load, the matrix in the composite laminates was damaged first, and the change of the load was not distinct at this time, which showed that the matrix damage had a little influence on the pin-load distribution. When the load continued to increase, the fiber in composite laminates was damaged, pin-load distribution slowly changed, the 1-row pin load gradually decreased, and the 3-row pin load gradually increased. When the structure was destroyed, pin-load distribution changed suddenly.

(2) Considering the cumulative damage of composite materials, pin-load distribution varied with the change of external load, but the change was small. Throughout the loading process, the pin-load distribution proportion showed that the 1,3-row pin was higher than the 2-row pin.

4.2 Laminate destruction process

The process of matrix damage development in a composite plate is shown in Figure 5. When the external load reached to 48kN, the laminates contacted with the 1- row bolts appeared bearing damage. With the load increasing, the matrix around the 3-row holes also underwent bearing damage, followed by the destruction of the matrix around the 2-row holes. Throughout the loading process, a number of the elements which matrix was damaged is the largest in the vicinity of 1-row holes, followed by the 3-row holes, and the least elements were damaged in the vicinity of the 2-row holes.

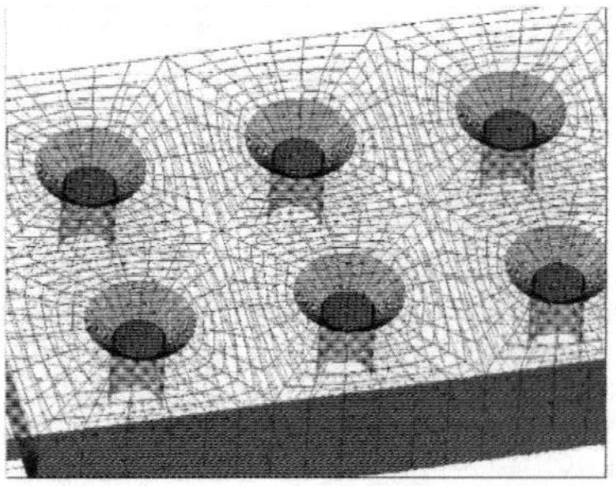
(a) 48 kN

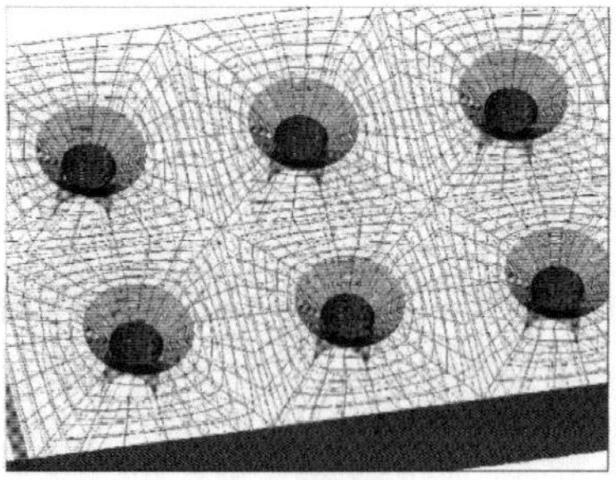
(b) 76 kN

Figure 5. Laminate matrix damage expansion process

The process of fiber damage development in composite plate is shown in Figure 6. A variety of failure modes appeared, including bearing damage of the matrix and fiber, and delamination as the external load was increased to 81kN. The fiber of laminated plate is crushed around the 1- row holes, that is, the part in contact with the bolts. As the load increased, the fiber around the 3-row holes also underwent bearing failure, followed by the fiber failure around the 2-row holes. When the load was increased to 115 kN, most of the fiber around the 1-row holes was damaged, losing load-carrying capacity. In the whole process of loading, the number of the elements damaged in fiber was the highest in the vicinity of 1- row holes, followed by 3-row holes, and the least fiber damage elements occurred near the 2-row holes.

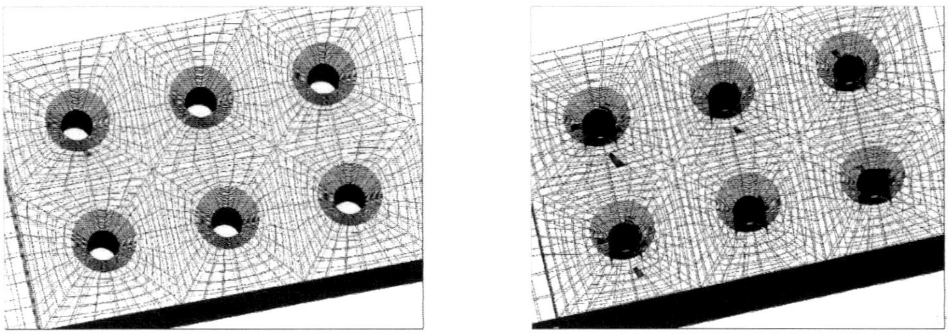

(a) 81 kN (b) 115 kN

Figure 6. Damage expansion process of composite laminates

4.3 Discussion

In Figure 4(a), the pin-load distribution does not change significantly with the increase of loading after the 1-row pins closest to the loading end overcome the friction force. The reason was that the matrix bearing damage occurred firstly around 1-row pins, then the bearing stiffness declined with the corresponding material properties degrading; The damage zone of the matrix was gradually increased with the loading increasing, so that the bearing stiffness continued to decline. However, as can be seen in Figure 5, a smaller number of damaged elements occurred in the stable phase of the pin-load distribution, it did not affect the pin-load redistribution.

When the fiber damage occurred, the loading on 1-row pins continued to decrease but did not cause a sharp change in pin load of each row. At the beginning of bearing damage, the damage zone was smaller and the stiffness degradation did not lead to rapid change of the pin load. On the one hand, the enlargement of the bearing damage zone reduced the overall bearing stiffness of the pin-hole. On the other hand, in the case of the bearing damage, the method of stiffness degradation was to degrade material stiffness abruptly then to amend and improve its stiffness. The bearing stiffness of the material had a tendency to improve after bearing damage. As a result of the combined effect of above-mentioned factors, the pin-hole bearing stiffness and load-bearing proportion decreased slowly. In addition, as the load increased, other pin-hole material gradually reduced stiffness, and it would further retard transferring process which the load of 1- row pins transfers to other rows pins. Another phenomenon observed from Figure 4(a) is that as the load-bearing proportion of one row of pins decreased, the load gradually shifted to 2-row pins, so that the load proportion of the middle-row pins increased slightly as the load increased, whereas the 3-row pins far from the loading end and 1-row pins, the loading was not significantly affected.

The bolt was made of metal, which was a ductile material and has remarkable plastic deformation before destruction. Therefore, in the multiple-bolted joints of metal materials, the hole which took a larger load firstly undergoes bearing plastic deformation, on the one hand, the stress concentration in the pin-hole was notably reduced; on the other hand, the load of each pin was redistributed so that the loading proportion tended to be uniform. At the same time, since the composite material was a typical brittle material and plastic deformation hardly occurs, which was not conducive to stress redistribution.

Consequently, the pin-load distribution would not change significantly with the change of external load.

5. Conclusion

Based on the theory of cumulative damage of composite laminates, this paper studied the problem of pin-load distribution of multiple-bolted in composite mechanical joints, and finally made the following conclusions:

(1) Considering the cumulative damage of composite material, the pin-load distribution varies with the change of external load, but the change is slight.

(2) Implementing the cumulative damage analysis technique of composite materials, it is found that pin-load distribution shows " Bolts at both ends bearing a greater load, the middle bolts bearing a smaller load ". The result is analogous to traditional analysis method that does not consider the effect of material damage on pin load.

(3) In order to consider the objective process of damage accumulation, iterative calculating is required in conducting FEM analysis of complicated structures, which leads to larger scale and more time-consuming computations by executing the strength calculation of composite joints. Under the condition that the accuracy of the two algorithms is similar, the algorithm which does not consider the material damage is more efficient and the calculation scale is small. To quickly determine the pin-load distribution in engineering applications, the effect of the cumulative damage of the material on the pin-load distribution should not be considered.

6. References

[1] Camanho P P and Matthews F L 1997 *Compos. Part A* **28A** 529-547.
[2] Askri R, Bois C, and Wargnier H 2016 *Procedia CIRP* **43** 292-296.
[3] Gray P J and McCarthy C T 2010 *Compos. Part II Engineering* **41** 317-325
[4] Álvaro O and Carlos S 2012 *Compos. Struct.* **94** 2110-2117.
[5] Cheng X Q, Wang S W, Zhang J, Huang W J, Cheng Y J and Zhang J K 2017 *Compos. Struct.* **160** 8-15
[6] Starikov R and Schon J 2001 *Compos.: Struct. B* 32 401-411.
[7] Tserpes K I, Labeas G and Papanikos P 2002 *Compos.: Part B* **33** 521-529
[8] Liu X D, Li Y Z, Su J and Ding R X 2014 *J. Aeronaut. Mater.* **34** 81-87
[9] Chang C R 1994 *Compos. Sci. Technol.* **50** 479-482.
[10] Nuismer R J and Labor J D 1978 *J. Compos. Mater.* **12** 49-60.
[11] Whitney J M and Nuismer R J 1974 *J. Compos. Mater.* **10** 253-265.
[12] Hashin Z 1980 *J. Appl. Mech.* **47** 329-335.
[13] Camanho P P, Davila C G and Demoura C G 2003 *J. Compos. Mater.* **37** 1415-1438
[14] Chang F K and Lessard L B 1991 *J. Compos. Mater.* **25** 2-43
[15] Tan S C and Perez J 1993 *J. Reinf. Plas. Compos.* **12** 1043-1057

Preparation and Properties of Cyanate/Epoxy-based Composite with Thermal Conductive Silica Particles

Xuemei He[1], Yuanming Chen[1, a], Kai Zhu[1], Wenjun Yang[1], Chong Wang[1], Wei He[1], Huaiwu Zhang[1], Yongqing Peng[2] and Kehua Ai[3]

[1]State Key Laboratory of Electronic Thin Films and Integrated Devices, University of Electronic Science and Technology of China, Chengdu 610054, China
[2] Netron Soft-Tech (Zhuhai) Co., LTD, Zhuhai 519040, China
[3]Sichuang Innoquick Electronics Co., Ltd, Suining 629000, China

E-mail: [a]ymchen@uestc.edu.cn

Abstract. Cyanate/epoxy (CE/EP) was added with surface-modified silica (SiO_2) particles to form a composite system. Microstructure of the composites was observed to find the dispersion effect of SiO_2 particles in the CE/EP matrix. Thermal conductivity, thermal stability and dielectric properties were also tested to investigate the effects of SiO_2 particles. CE/EP-based composite with 50 wt% SiO_2 exhibited thermal conductivity of 0.492 $W.m^{-1}.K^{-1}$, thermal decomposition temperature of 420 °C and dielectric constant of 4.1 at 1 MHz.

1. Introduction

Signal transmission at high speed calls for the low dielectric properties of polymer matrix since large dielectric loss can result in signal attenuation to limit signal integrity of electronic products [1, 2]. In addition, working reliability of electronic products could be challenged due to heat accumulation from high-power components [3]. Thus, good heat dissipation and low dielectric loss of polymer are needed to meet the requirement of high-efficiency working for electronic products.

Cyanate resin exhibits good heat stability and dielectric performance [4]. Cyanate/epoxy (CE/EP) blend from the coreaction of cyanate and epoxy monomers can also provide excellent performance due to the modification of network crosslinking density and the improvement of reaction extent form the individual functional groups. Thus, cyanate/epoxy is generally used as polymer matrix in composite system to further enhance its performance in the application of electronic products.

In this work, CE/EP was filled with modified silica (SiO_2) to enhance polymer performance. Thermal and dielectric properties were investigated to find out th1e effect of SiO_2 fraction on the composite system.

2. Experimental

SiO_2 particles were dispersed in γ-aminopropyltriethoxy silane (APS), ethanol and water solution with glacial acetic acid to regulate pH value of 3. The mixture was stirred to realize the adsorption of above silane agent for 1 h. After the centrifugation of above mixture at 2000 rpm for 15 min, the precipitate was then dried at 160 °C to completely remove the solvent. Thereby, surface-modified SiO_2 particles were obtained.

39 g CE and 8 g EP were blended to achieve a clear solution at 60 °C. Thereafter, modified SiO_2 particles with the fraction ranging from 10~50 wt% were added into the clear polymer solution under

stirring for 2 h at 120 °C. Above mixture was then transferred into polytetrafluoroethylene mould when SiO_2 particles were well dispersed in the polymer solution. Above mixture was then curried for 6 h at 200 °C after air bubbles were avoided during the vacuumizing for 1 h. Thus, CE/EP-based composites with SiO_2 particles were finally obtained. CE/EP and CE/EP-based composites were shown in Fig.1.

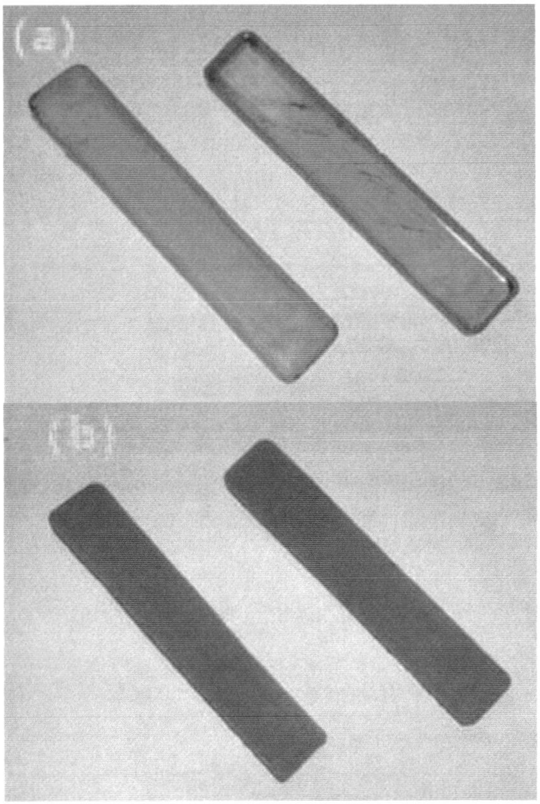

Figure 1. Optical images of CE/EP (a) and CE/EP-based composites (b)

Fourier-transform infrared (FT-IR) spectroscopy was used to determinate chemical composition of SiO_2 particles. Cross sections of SiO_2/CE/EP composites were observed using a scanning electron microscope (SEM). A thermal conductive measurement was used to evaluate thermal conductivity of SiO_2/CE/EP composites and CE/EP matrix. Thermo gravimetric analysis (TGA) of the composites was performed from room temperature to 650 °C at a heating rate of 10 °C /min under nitrogen atmosphere. Measurements of dielectric properties were performed at the frequency of 1 MHz on a broadband dielectric spectrometer.

3. Results and Discussion

3.1. Chemical Composition of SiO_2

Silane coupling agents with silane functionalization at the surface of inorganic particles can improve wettability and dispersion of inorganic particles in the polymer matrix. SiO_2 particles were treated by APS to enhnace their dispersion due to the decrease ofinterface energy. The chemical compositions of SiO_2 particles were characterized with FTIR spectra, as shwon in Fig. 2. The peaks at 1085.8cm^{-1} and 791.5cm^{-1} was assigned to the characteristic absorption of SiO_2 particles. The absorption bands at 1090 cm^{-1} and 1220 cm^{-1} corresponded to stretching vibration of siloxane Si–O–Si and flexible vibration of Si–C in SiO_2 particles with surface treatment of APS (SiO_2-APS). Thus, the interface of SiO_2 particles has adsorbed by the oligomeric silsesquioxane after the reaction of APS.

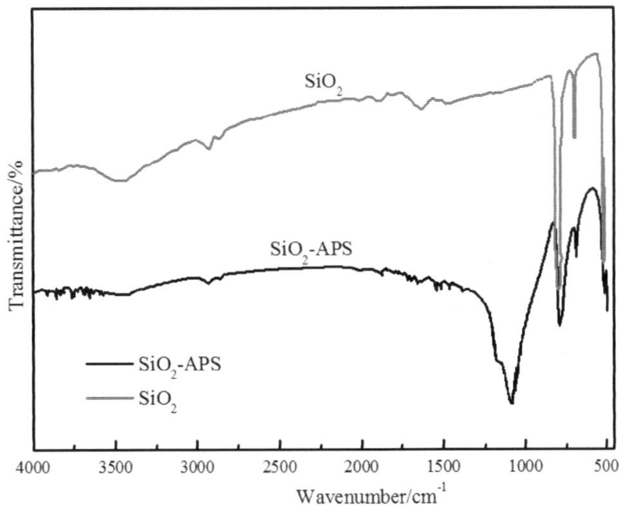

Figure 2. FTIR spectra of SiO_2 particles before and after treatement

3.2. Microstructure of Composites.

Silane coupling agents with both inorganic and organic groups can absorb at the interface of inorganic fillers to create a wetting bridge to well blend with organic polymer [2, 5]. Fig. 3 presented the microstructure of SiO_2/CE/EP composites. SiO_2 particles were well dispersedly in the CE/EP matrix when the fraction of SiO_2 particles reached 10 wt% since SiO_2 particles with the absorption of silane coupling agent could provide stable surface for SiO_2 particles to break their mutual aggregation. However, contact chance between SiO_2 particles increased when 50 wt% SiO_2 was added into the CE/EP matrix. This could be due to gradual saturation of SiO_2 in polymermatrix.

Figure 3. SEM micrographs of the composites as a function of particle fractions: (a) 10 wt%; (b) 20 wt%; (c) 30 wt%; (d) 40 wt%; (e) 50 wt%

3.3. Thermal Conductivity of Composites

Fig. 3 displayed the thermal conductivity of the composites with different fractions of SiO_2 filler. CE/EP matrix exhibited very low thermal conductivity of 0.225 $W.m^{-1}.K^{-1}$. However, the addition of SiO_2 particles into CE/EP matrix led to an obvious increase of thermal conductivity for the composite system. The thermal conductivity of the composite reached to 0.312 $W.m^{-1}.K^{-1}$ when 10 wt% SiO_2 particles were added into the polymer matrix. In addition, the thermal conductivity had a jump in the condition of the composites with more than 30 wt% SiO_2 particles. Especially, the thermal conductivity of the composite with 50 wt% SiO_2 particles dramatically increased to 0.492 $W.m^{-1}.K^{-1}$, 0.267 $W.m^{-1}.K^{-1}$ higher than the CE/EP matrix without filler. The phenomenon on the change of thermal conductivity for the composited could be attributed to inorganic higher thermal conductivity of SiO_2 particles and the formation of conductive network from more mutual contact between SiO_2 particles in high fraction [6].

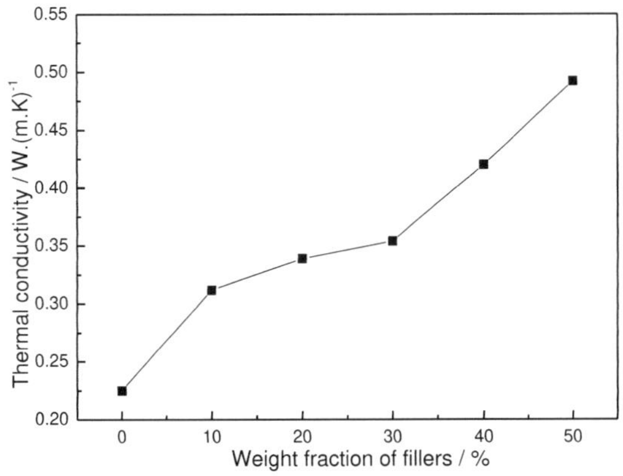

Figure 4. Thermal conductivity of CE/EP-based composite with different fractions of SiO_2 particles

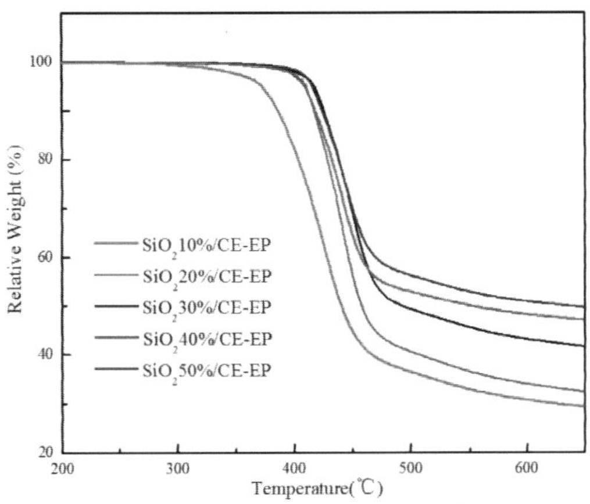

Figure 5. Thermogravimetric profiles of composites with different fractions of SiO_2 particles

3.4. Thermal Stability of Composites

TGA of CE/EP-based composites with different filler contents was measured to investigate the effects of SiO_2 particles on thermal stability, as indicated in Fig. 5. The composites exhibited an increase of thermal decomposition temperature from 300 °C to 420 °C when the fraction of SiO_2 particles increased from 10 wt% to 50 wt%. In addition, the char residue of the composites increased after thermal decomposition. CE/EP-based composite with 50 wt% SiO_2 particles had 51% char residue, 32% bigger than the case for the composite with 10 wt% SiO_2 particles. Generally, high concentration of SiO_2 particles as the obstacle could influence the copolymerization of long network chain for polymer to weaken the thermal stability of the composite system. However, SiO_2 particles performed not only the effect of heat absorption but also heat transmission to act probably as a barrier for oxygen diffusion so that the composites with high fraction of SiO_2 particles exhibited good thermal stability [7, 8].

3.5. Dielectric Constant of Composites

CE/EP with low dielectric constant was usually employed to avoid signal distortion for electric devices with high-speed signal transmission [2, 9]. However, the defect of low thermal conductivity for polymeric CE/EP matrix resulted in the limitation on its application of high-power electric devices. Thus, SiO_2 particles with low dielectric constant and high thermal conductivity were added into CE/EP matrix in order to improve its performance. Fig. 5 presented dielectric constant of the composites with SiO_2 particles at 1 MHz. Higher concentrations of functionalized particles resulted in higher dielectric constant of composites. At 50 wt% fraction in the CE/EP-based composite, surface-modified SiO_2 particles provided the dielectric constant of 4.1 at 1 MHz, 0.9 higher than that of CE/EP matrix. The addition of SiO_2 particles in the composite system resulted in a small increase of dielectric constant due to the small difference of dielectric constant between SiO_2 filler and CE/EP matrix [10]. In this way, SiO_2 particles were a potential filler to modify the thermal and electric properties of polymeric CE/EP matrix.

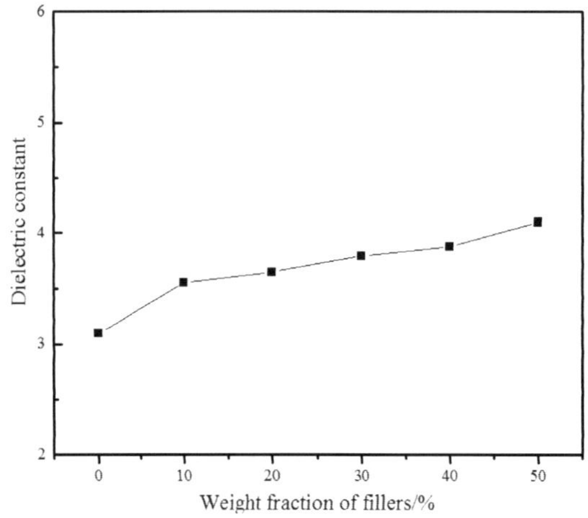

Figure 6. Dielectric constant of composites with different fractions of SiO_2 particles at 1 MHz

4. Conclusions

In this study, SiO_2 particles with surface modification of γ-aminopropyltriethoxy silane were used as the inorganic filler to enhance the performance of polymeric CE/EP. SiO_2 particles were well dispersedly in the CE/EP matrix after their surface modification. Thermal conductivity of CE/EP-based composite increased to 0.492 $W.m^{-1}.K^{-1}$ when the fraction of SiO_2 particles reached 50 wt%. Higher fraction of SiO_2 particles led to better thermal stability of the composite system. SiO_2 particles with low dielectric constant caused limited influence on the dielectric constant of the composite system.

5. Acknowledgements

The authors gratefully acknowledge the support of the projects of National Natural Science Foundation of China (No.61240055 and No.61604034), and we also express our sincere thanks to project of science and technology planning of Guangdong Province China (No.2016B090918095).

6. References

[1] C. Wang, N. Wen, G. Zhou, S. Wang, W. He, X. Su and Y. Hu: Applied Surface Science 422 (2017), p. 738-744.
[2] Y. Chen, X. He, Y. Wu, X. Gao, J. Wang, W.He, V. V. Silberschmidt and H. Xu: Polymer Composites 37 (2016), p. 3033-3041.
[3] Y. Chen, S. Wang, X. He, W. He, V. V. Silberschmidt and Z. Tan: Circuit World, 41 (2015), p. 55-60.
[4] T. Lu, G. Liang and Z. Guo: Journal of Applied Polymer Science 101 (2006): p. 3652-3658.
[5] S. Kuo and F. Chang: Progress in Polymer Science, 36 (2011), p. 1649-1696.
[6] Y. Chen, X. Gao, J. Wang, W. He, V. V. Silberschmidt, S. Wang, Z. Tao and H. X: Journal of Applied Polymer Science, 132 (2015), p. 41889.
[7] S. Liang, X. Yang, Xi. Fang, W. D. Cook, G. A. Thouas and Q. Che: Biomaterials 32 (2011), p. 8486-8496.
[8] Y. Chen, W. He, G. Zhou, Y. Hu, S. Wang and Z. Tao: Polymer International 63 (2014), p. 546-551.

[9] L. Liu, F. Lv, P. Li, L. Ding, W. Tong, P. K. Chu and Y. Zhang: Composites Part A: Applied Science and Manufacturing 84 (2016), p. 292-298.
[10] N. H. Hieu: International Journal of Materials Science and Engineering 5 (2017), p. 1-15.

IOP Conf. Series: Materials Science and Engineering **422** (2018) 012004 doi:10.1088/1757-899X/422/1/012004

Finite Element Method for Predicting the Behavior of Sandwich Structure Luggage Floor of Passenger Cars

Alok Raj[1, a], **Meenakshi Akkiraju**[2, b], **Balaji Pillai**[3, c], **Naresh Kulkarni**[4, d], **Rathnakara Somashekara**[5, e]

[1, 2,3,4,5] Mercedes Benz Research and Development India, 560066, Bangalore, India

E-mail: [a]alok.raj@daimler.com, [b]meenakshi.akkiraju@daimler.com, [c]balaji.pillai@daimler.com, [d]naresh.kulkarni@daimler.com, [e]rathnakara.somashekara@daimler.com,

Abstract. Sandwich structures are gaining more importance in automobile industry due to its low weight, high flexural stiffness, and corrosion free properties. The composite structure with paper core sandwiched between two Polypropylene Glass-Filled (PP-GF) faces are being utilized for load floor of luggage compartments in passenger cars. To design, develop, and optimize the luggage floors, reliable Finite element (FE) methods are essential.

This study deals with developing a finite-element method to predict the behavior of the luggage floor of a typical Sedan car during ball impact test and static stiffness test. FE simulations are performed using LS-Dyna explicit code. Honeycomb core is modelled using solid elements and Modified honeycomb material model in LS-Dyna is assigned to the honeycomb structure. The two faces, bottom and top, are modelled using shell elements and modified-piecewise linear plasticity material model is assigned to them, respectively.

Experimental investigations are done for quasi-static four point bend test. FE simulations are performed for quasi-static four point bend test and validated with experimental results. Based on validated FE method, the ball impact test and static push test are simulated. Experiments are performed for the ball impact test and static push test. The results are compared and correlation between experiment and FE simulation is studied by comparing the deformation pattern, forces, and displacements. This developed FE method is for predicting the behavior of the luggage floor.

1. Introduction

Honeycomb sandwich composites must have high bending stiffness to be used as luggage floors in automobiles (Figure 1). Honeycomb composite structures, which comprises of three layers viz., two face sheet and a core material, have good bending stiffness. Over the past few decades, various researchers all around the world have shown keen interest to get mechanical properties experimentally and tried to develop modelling techniques to predict its behavior. Heimbs et al. [1, 2] predicted the structural response of honeycomb sandwich material and focused on modelling and numerical analysis. Seemann and Krause [3] analyzed various approaches of modelling Nomex honeycomb and suggested orthotropic elasto-perfectly plastic material model to be perfect. Shipsha and Zenkert [4] did detailed experimental investigation showing core crushing damage and compression after effect for composite sandwich panels. The Honeycomb structure studied in this paper consists of hexagonal paper honeycomb sandwiched between anisotropic polypropylene glass fibers (PP-GF).

In the literature, many of the simulations are done using detailed modelling of honeycomb cores. A Finite Element method with simpler approach of modelling honeycomb core needs to be developed to enable the use of FE method for faster design and development of sandwich structures in automobiles. The current work aims to use the shell-solid-shell concept to develop a simple and reliable FE modelling method to predict the behavior of the sandwich structures. Experimental studies are conducted to validate the proposed FE method.

(a)

(b)

Figure 1. (a) Rear view of car with luggage compartment (b) Detailed view of luggage floor

2. Experimental Study

The experimental results are required to validate the behavior of the sandwich composites in proposed FE method. Experiments are performed under room temperature at both static and dynamic loading conditions. The following experiments are conducted.

2.1. Static Four Point Bend Test

In accordance with DIN standards 53293 [5], the flexure experiments are performed on various specimens. The specimen dimensions are tabulated in table 1 and test specifications are shown in table 2. The specimens are subjected to displacement-controlled loading. The test set-up is shown in figure 2(a) and a typical force Vs displacement plot is shown in figure 2(b).

Table 1. Test specimen

Specimen		Measurement
Length	[mm]	380
Width	[mm]	60
Thickness	[mm]	10

Table 2. Test specification

4-Point bend test		Description
Test Standard/Work Instruction		DIN 53293
Test control mode		Stroke
Rate of Loading	[mm/min]	3
Load span Length	[mm]	150
Support Span Length	[mm]	300
Diameter of support and loading roller	[mm]	10

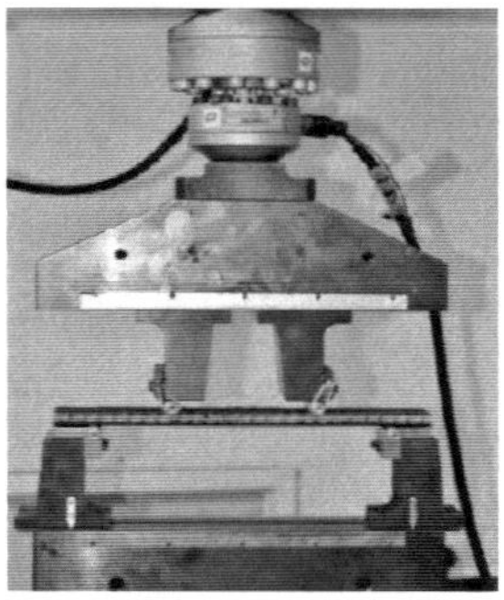

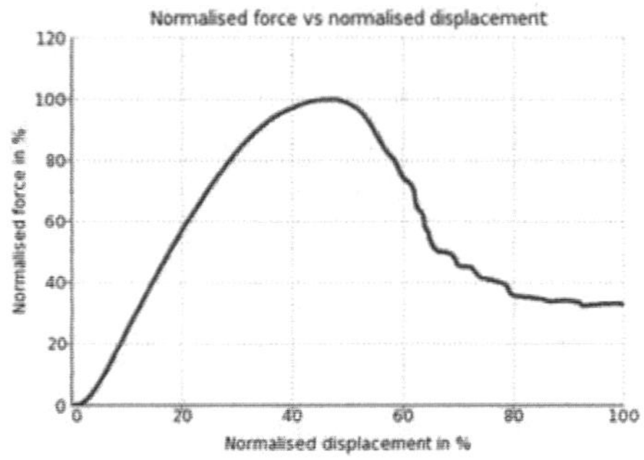

(a) (b)

Figure 2. (a) Test Set-up (b) Load-displacement plot

2.1.1. Results. The normalized force vs displacement curves are plotted as in Figure 2(b).

2.2. Static Push Test.

A static test is performed on the load floor using a rectangular loading device. The load is held constant for 5sec and then unloaded. Maximum deflection and permanent deformation are checked,

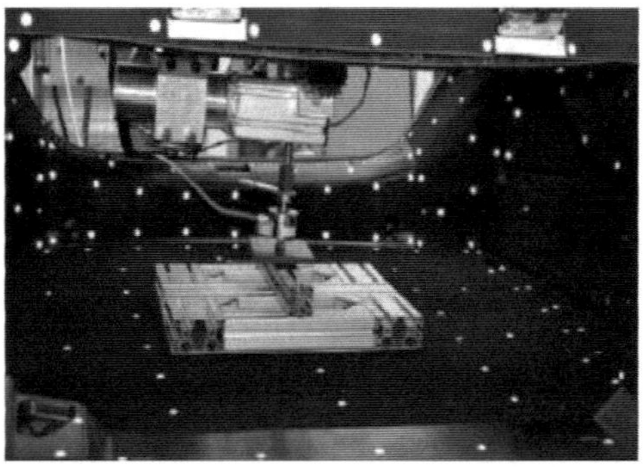

Figure 3. Static push test set-up

2.2.1. Results. The maximum vertical deflection measured. After unloading, permanent deflection experienced by floor is also measured.

2.3. Dynamic Ball Drop Test

2.3.1. Test Set up. To get dynamic response of the part, a rigid ball with a specific mass is raised to a specific height as shown in figure 4(a). The Impactor freely falls due to gravity onto the luggage floor. The deflection of floor is measured using a transducer positioned in touch condition beneath the floor in the spare-wheel tub as shown in figure 4(b). An accelerometer mounted on top of ball records the acceleration. The set-up of ball drop test is shown in Figure 4(c).

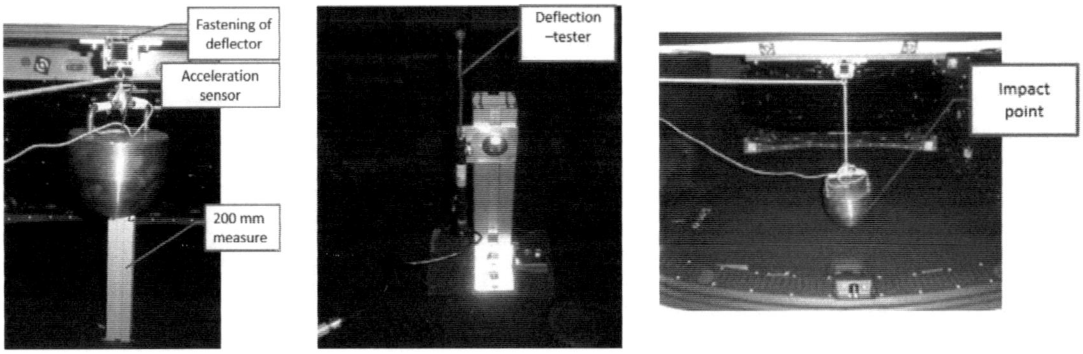

(a) (b) (c)

Figure 4. Test set-up for dynamic ball drop test

2.3.2. Results. The maximum vertical deflection measured. The Permanent deflection experienced by floor is also measured.

3. Numerical Study

3.1. Material Modelling

The luggage floor consists of honeycomb structured paper and PP-GF face sheet. The mechanical properties of honeycomb paper depends upon direction of loading. Six normal and shear stress strain curves [6] are used to define the orthotropic properties of solid honeycomb core. The FE method under investigation proposes to use simple solid model instead of actual honeycomb geometry to avoid complexity. But representing honeycomb structures with simple solid layer would result in subsequent increase in strength and stiffness. Hence the material stress-strain curves need to be scaled accordingly. To obtain the scaling factor for curves, an empirical scaling approach is utilized.

3.1.1. Empirical Scaling of Material Properties for Finite Element Simulations. The purpose of this method is to scale the stress-strain curves of honeycomb core material when replaced by solid element. A FE model is prepared taking actual geometrical dimensions of honeycomb. Figure 5 demonstrates the actual FE model of honeycomb and solid structure.

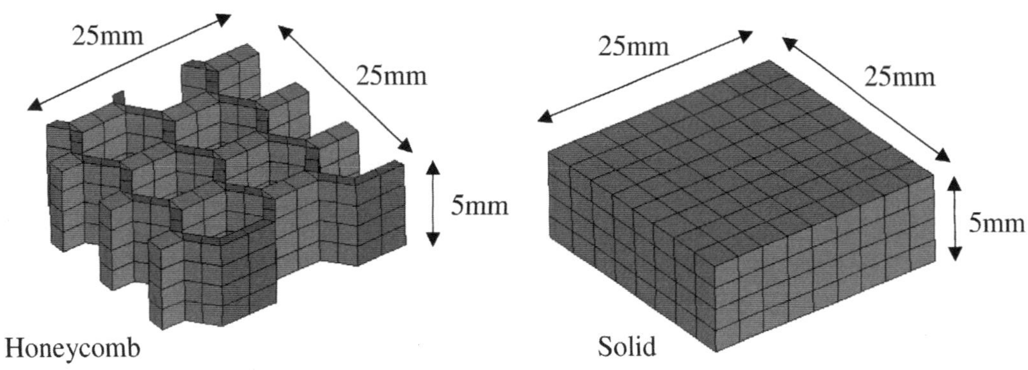

(a) (b)

Figure 5. (a) Actual honeycomb FE model (b) Simplified solid FE model

The FE mesh of hexagonal geometry is prepared by taking measurements from the sample. A compression load is applied to actual geometry and simple solid model. Figure 6 illustrates the force direction and FE model at compressed state.

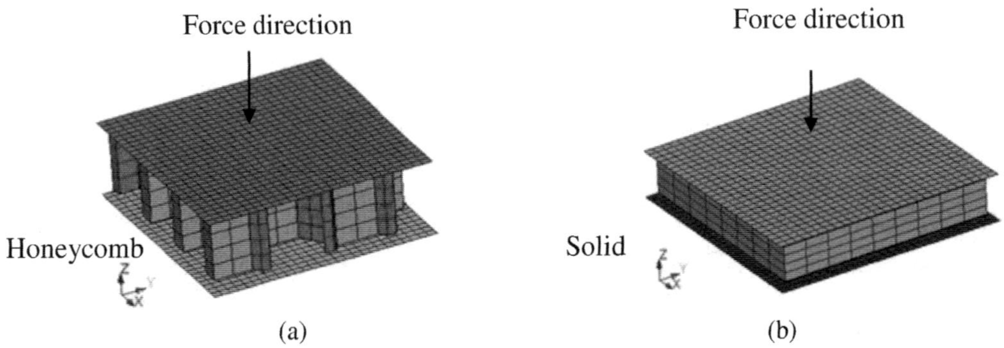

Figure 6. (a) Compressed honeycomb FE model (b) Compressed simplified solid FE model

The maximum reaction force offered by actual honeycomb structure (Factual) normalized is 100% whereas by simple solid structure (Fsolid) is 416%. Based on force comparison, an empirical scaling factor is obtained. Emperical scaling factor (Factual / Fsolid) = 0.24. The stress-strain curve defining the off-axis properties of honeycomb paper is scaled by this factor. The scaled stress-strain curves are used to simulate the compression test again with solid elements. The results are compared with the actual honeycomb geometry and forces are found to be similar. Similar approach is taken by Kayran and Aydineak [7] to replace aluminium honeycomb core with equivalent continuum core. A good level of correlation between actual and scaled model is observed. Table 3 exhibits the resultant force and corresponding displacement for actual honeycomb and simple solid model with and without tuning the stress-strain curves.

Table 3. Resistance force offered by original and scaled materials

Material	Normalized Force (in %)
Solid with Honeycomb structure	100%
Simple Solid structure with original material	416%
Simple Solid structure with scaled material	102%

3.2. FE Modelling.

Replication of the actual geometry of honeycomb structure is complex and computationally expensive. This limits the usage of proposed FEM in digital development of sandwich parts. Hence, a simple solid element based FE modelling technique is proposed, where honeycomb structures are represented as hexagonal solid elements with element size of 3mm. Commercially available LS-Dyna Explicit code is used for simulation. Modified honeycomb material with scaled stress-strain properties are used. The PP-GF material on top and bottom of the paper is modelled as shell elements and Modified-piecewise linear plasticity material [8] is assigned to it. Solid honeycomb and shell face of PP-GF are in contact through nodal merge. A contact definition of Automatic Single Surface is used. Delamination modelling is not considered, as the real life use cases under consideration of present study does not load the honeycomb part to the extent of considerable delamination.

3.3. Simulations

3.3.1. Four Point Bending Simulation. Specimen rests on two supports as shown in figure 2(a) and dimensions are similar to as mentioned in Table 1. For the two loading devices, all rotational degrees of freedom are constrained and allowed to move vertically only. A displacement-controlled quasi-static movement is given to the rollers placed with touch condition on top of the specimen. In the simulation, the reaction force is measured from both the two loading devices. Figure 7 (a) & (b) shows the simulation view and corresponding variation in total force with imposed displacement.

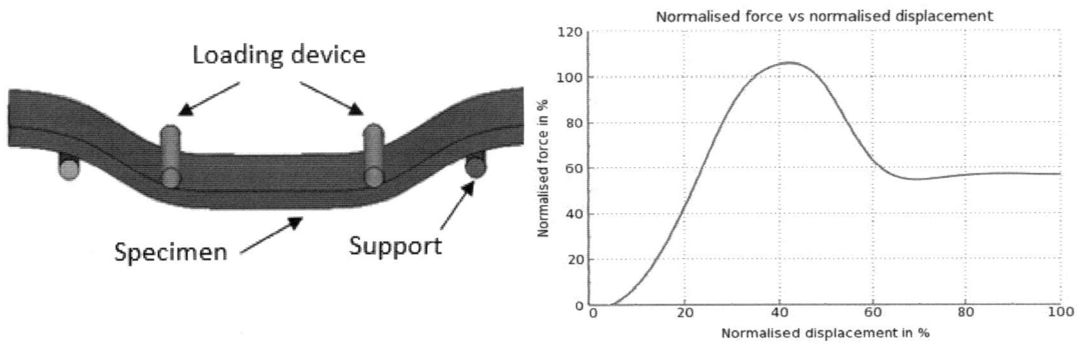

(a) (b)

Figure 7. (a) Simulation view (b) Force-displacement curve

3.3.2. Push Simulation. Simulation is done similar to the experiment. Force is applied quasi-statically and held for sufficient time for load stabilization, followed by unloading. Maximum and permanent deflections are measured, respectively. Figure 8 (a) & (b) shows the initial and deformed state of floor in push test simulation.

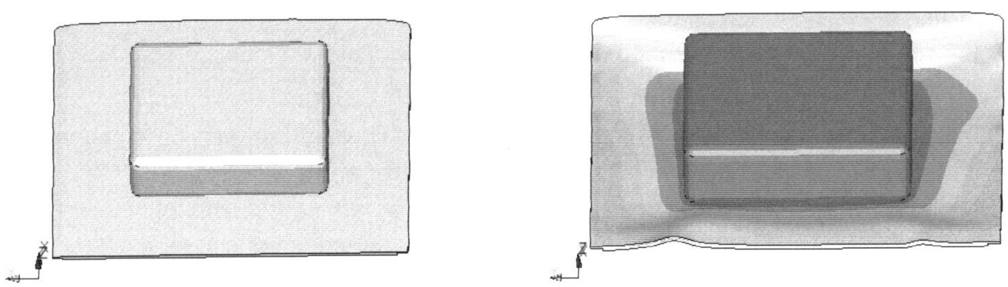

(a) (b)

Figure 8. (a) Initial state of luggage floor (b) Deformed state of luggage floor

3.3.3. Dynamic Ball Drop Simulation. Simulation is done similar to the experiment. The reaction force is measured on the impactor. A local indentation mark is clearly visible in simulation same as seen in the test. The maximum force and deflection are noted. The permanent deflection is also noted. Figure 9 (a) & (b) shows the initial and deformed state of floor in ball drop simulation.

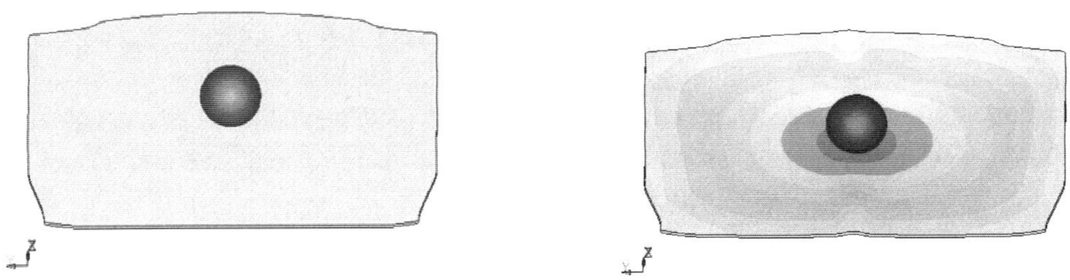

(a) (b)

Figure 9. (a) Initial stage of luggage floor (b) Deformed state of luggage floor

4. Validation of FE with Experimental Results

The proposed FE method of composite material with honeycomb core sandwiched between two face sheets is validated by using the four-point bend experiment results. Simulation results are normalized for force and deflection from the experimental results and are compared. Comparison is described in Table 4. The deformation pattern is observed to be similar.

Based on the aforementioned validation, the FE model is used for static push and ball impact. Simulation results for normalized force and normalized deflection are compared in Table 4 for static push test and the ball impact test. Good correlation is observed which suggests the applicability of the proposed FE method for predictive engineering. Minor variation in force could be attributed to the failure and the delamination of face sheet from the core, which is not expected in the simulation. Experiment and simulation results of the all the studies have been tabulated below (Table 4).

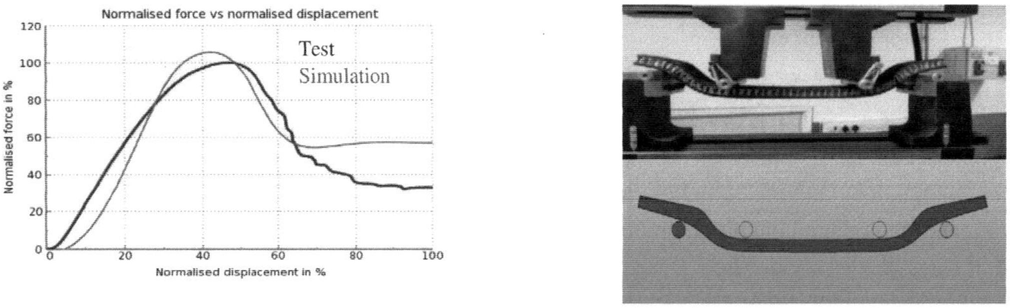

(a) (b)

Figure 10. (a) F-D curve of simulation overlaid with curve obtained from experiment (b) Simulation and Test specimen at deformed state

Table 4. Comparison b/w test and simulation for four point flexural test, push test, and impact test

	Flexural Test vs Simulation		Push Test vs Simulation		Impact Test vs simulation	
	Force (N)	Deflection (mm)	Force (N)	Deflection (mm)	Force (N)	Deflection (mm)
Experiment	100 %	43.3 %	100 %	102.5 %	100 %	93.5 %
Simulation	105.6 %	41.6 %	100 %	97.5 %	118.7 %	95 %

5. Conclusion and Future Scope

Finite Element Method using simple shell-solid-shell with scaled material properties are proposed to model sandwich structures. A good level of correlation for four point bend test, push test, and ball drop test, with the proposed FE method is achieved, respectively. This proposed FEM is suggested to be used for digital design and development of the luggage floor or similar components made of honeycomb structures during the early phase of design. This could reduce the number of hardware tests and get the perfect designs during the initial phase itself. This in turn could lead to reduction in costs and turnaround time. Delamination prediction, element size sensitivity, and more test cases, need to be experimentally validated to improve the method. This is suggested as future work.

6. Acknowledgements

The authors gratefully acknowledge the support provided by the Mercedes-Benz Research and Development India (MBRDI) and Daimler AG (DAG), for the research work leading to this paper. Authors would like to express their gratitude to Mr. Parthiv Shah, Head of the Department, Safety, MBRDI who rendered his valuable support during the research period. Authors expresses their gratitude to colleagues of RD/KIA (Interiors) department at Daimler AG for their valuable support during the research.

7. References

[1] Heimbs, S., Mehrens, T., Middendorf, P., Maier, M. and Schumacher, A., 2006. Numerical determination of the nonlinear effective mechanical properties of folded core structures for aircraft sandwich panels. In 6th European LS DYNA User's Conference.

[2] Heimbs, S., Vogt, D., Hartnack, R., Schlattmann, J. and Maier, M., 2008. Numerical simulation of aircraft interior components under crash loads. International journal of crashworthiness, 13(5), pp.511-521.

[3] Seemann, R. and Krause, D., 2017. Numerical modelling of Nomex honeycomb sandwich cores at meso-scale level. Composite Structures, 159, pp.702-718.

[4] Shipsha, A. and Zenkert, D., 2005. Compression-after-impact strength of sandwich panels with core crushing damage. Applied Composite Materials, 12(3), pp.149-164.

[5] Feldhusen, J., Torsakul, S., Brezing, A. and Krishnamoorthy, S., 2008. Numerical Modeling and Experimental Investigation of the Failure Modes of the Cellular Foam Sandwich Structures. Journal of Metals, Materials and Minerals, 18(2), pp.111-115.

[6] Heimbs, S., Middendorf, P. and Maier, M., 2006, June. Honeycomb sandwich material modeling for dynamic simulations of aircraft interior components. In Proceedings of the 9th international LS-DYNA users conference. Dearborn (MI).

[7] Kayran, A. and Aydıncak, İ., Assessment of Effective Elastic Properties of Honeycomb Cores by Finite Element Analysis of Sandwich Panels.

[8] L.S DYNA Keyword user manual. Version 8, 2008.

IOP Conf. Series: Materials Science and Engineering **422** (2018) 012005 doi:10.1088/1757-899X/422/1/012005

Macro-mechanical Analysis on the Variation of Fibre Orientation in a Composite UAV Landing Gear

Harsh Raj Chauhan[1,a], **Vikas Rastogi**[1,b], **Atul Kumar Agarwal**[1,c]

[1]Mechanical Engineering, Delhi Technological University, Delhi, India

Email: [a]harshchauhan1995@gmail.com,[b]rastogivikas@gmail.com, [c]atulkumaragarwal@gmail.com

Abstract: Composite materials are increasingly being relied upon for their light weight and high strength to weight ratios and form the backbone of the aerospace industry, which follows a Systems Engineering approach. This paper is an attempt to manifest such an approach at a relatively smaller scale for a composite UAV landing gear. The landing gear acts as a support structure for the fuselage during take-offs and landings and hence needs to be carefully designed, analyzed and manufactured. The strength and weight of a landing gear has to be optimized so as to adhere to the overall aerodynamics and structural integrity of the aerial system. This study is initiated with the design concept of the landing gear and the state of load being acted upon followed by appropriate composite material selection. An approximate number of layers are decided for the composite vacuum layup of the landing gear through iterative experimentation. The fibre orientation of the composite layers is instrumental in gauging the macro-mechanical performance of the composite, thus being the focal point of this study. To execute this, the extensional, coupling and bending stiffness matrices are computed in a mathematical model in MATLAB for different fibre orientations in response to the elastic constants of the fibre and matrix. The global and local strains and curvatures are calculated for a load case and the final combination is down-selected. Corresponding to the down-selected combination, the landing gear is finally fabricated, using Vacuum bag moulding.

Nomenclature:

E1 =*Longitudinal Young's Modulus*
E2 =*Transverse Young's Modulus*
v12 =*Major Poisson's Ratio*
v21 =*Minor Poisson's Ratio*
G12 =*In-plane shear modulus*
[A] = *Extensional Stiffness Matrix*
[B] = *Coupling Stiffness Matrix*
[D] = *Bending Stiffness Matrix*
Nx, Ny = *Normal force per unit length*
Nxy = *Shear force per unit length*
Mx, My = *Bending moments per unit length*
Mxy = *Twisting moments per unit length*
$[Q]$ = *Reduced stiffness matrix*
$\overline{[Q]}$ =*Transformed reduced stiffness matrix*
e_x, e_y, e_{xy} = *Mid-plane strains*

Kx, Ky, Kxy = *Mid-plane curvatures*
[T] =*Transformation matrix*
[R] = *Reuter matrix*
(1,2) = *Local axes denomination*
(x,y) = *Global axes denomination*

1. Introduction

Systems Engineering, which supplements the design and development of unmanned aerial vehicles follow an inter-disciplinary approach to tackle complex systems such as this one. The landing gear of an unmanned aerial vehicle (UAV) functions as a shock absorber for the fuselage of the UAV, shielding the payload. The dynamic loads acting on the UAV due to sudden gusts and abrupt maneuvers are fully absorbed by the landing gear. It therefore demands a meticulous structural design of the landing gear of the UAV. With the advent of hybrid materials like composites and smart materials, it calls for careful selection of materials and their macro-mechanical analyses of resulting inner fiber and matrix to reach a design with a high technology readiness level.

2. Statement of Objectives

The design objectives for the composite landing gear are bifurcated into high-level and low-level objectives based on their functionality and dependence. The list of objectives in Table 1 presents the objectives kept in mind while approaching the design problem.

Table 1. Statement of objectives for the design of a composite landing gear

S.no	Parameter/Objective	Value/Range
1	Weight	5-7% of UAV weight
2	Static Load on the landing gear	85-90 N
3	Dynamic Load on the landing gear	150-170 N
4	Factor of safety	1.5
5	Strength/Weight %	130-180
6	Material characteristics	CF fabric/Epoxy matrix
7	Technology Readiness Level	5-7

3. Engineering Design

The landing gear is designed to tested for the static load of 85 N and an assumed dynamic load of 150N [1], it may take during the flight along with suitable safety factor, and several materials are qualitatively judged on a number of parameters. Carbon-fibre is finally selected among a list of other metals and composites. Different grades of carbon fibres are tested ranging from 200-400 gsm through a mathematical model and 400 gsm is finally selected.

Figure 1. CAD model of the landing gear

Figure 2. Raw 3D CAD of the UAV

4. Macro-mechanical Analysis

The macro-mechanical analysis of a composite lamina is based on the consideration that the lamina is homogeneous and the average properties of the individual layers are calculated [2]. The stiffness matrices are calculated based on the engineering constants of uni-directional lamina (both fiber and matrix) using a mathematical model.

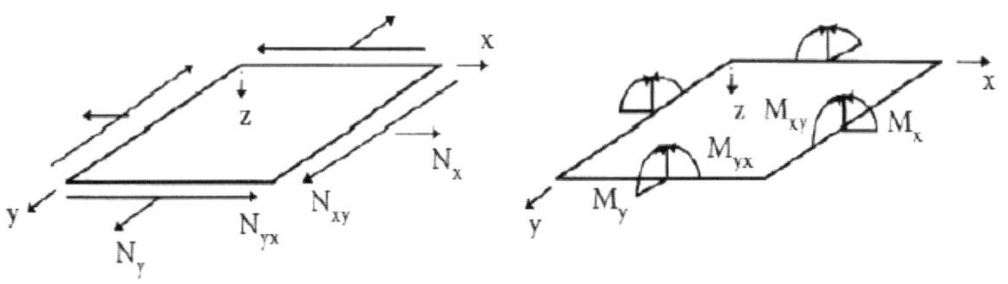

Figure 3. Forces & moments acting on a composite laminate [2]

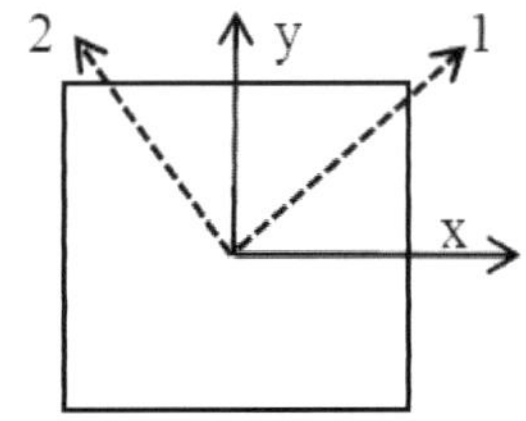

Figure 4. Local & Global axes representation of a lamina [3]

4.1. Mathematical Model.

The mathematical model takes as input, the elastic moduli of the fiber and the matrix of the composite, as specified in [4] to advance to the combined elastic moduli, *E1, E2, v12* and *G12* to finally calculate the reduced stiffness matrix *[Q]* for each ply of the composite using the following governing equations. (Refer Eq.1 to Eq.4)

$$Q11 = E1 / (1 - v12 * v21) \quad (1)$$
$$Q12 = E2 * v12 / (1 - v12 * v21) \quad (2)$$
$$Q22 = E2 / (1 - v12 * v21) \quad (3)$$
$$Q66 = G12 \quad (4)$$

The lamina in consideration is taken as purely orthotropic [5], so the other stiffness coefficients result in value of 0. Based on the reduced stiffness coefficients and restricting the number of plies, the fiber orientation of each ply is arbitrarily taken, considering the ease of fabrication. The number of plies are kept constant at 4, due to restricted weight constraint and the number of test cases are increased to 6 with the following layup.

Layup-1	*Layup-2*	*Layup-3*	*Layup-4*	*Layup-5*
0	0	0	0	0
90	90	45	45	-45
0	90	0	45	45
90	0	45	0	0

Figure 5. Layups with 4 plies each, with their fibre orientations

Each of the layups are separately taken and the transformed reduced stiffness matrix $[\overline{Q}]$ are calculated for each ply. Several assumptions were made in the process, for uniformity in the mathematical model and ease of fabrication.

Assumption 1 : The thickness of the plies were estimated, assuming uniform thickness of matrix(epoxy) and the fibre.

Assumption 2 : The mid-plane of the layup was assumed as origin for calculation of the coordinates of each ply.

Using the stiffness coefficients $[Q_{ij}]$ and the location of each ply, the extensional *[A]* , coupling *[B]* and bending *[D]* stiffness matrices are calculated [1], (Refer Eq.5 to Eq.7)

$$A_{ij} = \Sigma[\overline{Q_{ij}}]_k * (h_k - h_{k-1}) \text{ where } i=1,2,6 \text{ ; } j=1,2,6 \text{ ; } h_k= \text{coordinate of kth ply} \quad (5)$$
$$B_{ij} = \Sigma[\overline{Q_{ij}}]_k * (h_k^2 - h_{k-1}^2) \quad (6)$$
$$D_{ij} = \Sigma[\overline{Q_{ij}}]_k * (h_k^3 - h_{k-1}^3) \quad (7)$$

Corresponding to the obtained matrices *[A], [B]* and *[D],* the required forces and moments are applied to the final configuration of the layup and the mid-plane strains and curvatures are found by solving 6 simultaneous equations.(Refer to Eq.8)

$$\begin{bmatrix} N_x \\ N_y \\ N_{xy} \\ M_x \\ M_y \\ M_{xy} \end{bmatrix} = \begin{bmatrix} A_{11} & A_{12} & A_{16} & B_{11} & B_{12} & B_{16} \\ A_{12} & A_{22} & A_{26} & B_{12} & B_{22} & B_{26} \\ A_{16} & A_{26} & A_{66} & B_{16} & B_{26} & B_{66} \\ B_{11} & B_{12} & B_{16} & D_{11} & D_{12} & D_{16} \\ B_{12} & B_{22} & B_{26} & D_{12} & D_{22} & D_{26} \\ B_{16} & B_{26} & B_{66} & D_{16} & D_{26} & D_{66} \end{bmatrix} \begin{bmatrix} e_x \\ e_y \\ e_{xy} \\ K_x \\ K_y \\ K_{xy} \end{bmatrix} \quad (8)$$

The local and global strains are also evaluated using the transformation operator $[R][T][R]^{-1}$. Finally, 6 variables are calculated through the mathematical model, 3 each for midplane strains and mid-plane curvatures.

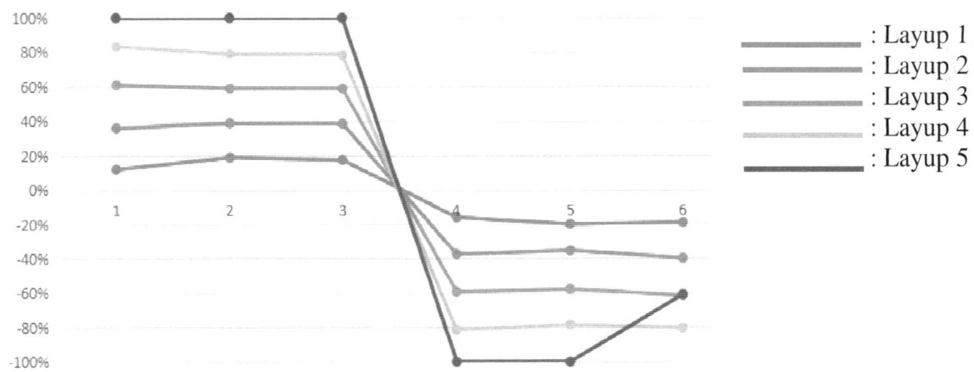

Figure 6. Percentage variation of the values of 6 variables for 5 layups in the mathematical model

4.2. Finite Element Method.

To analyze the accuracy of the results obtained from the mathematical model, a finite element simulation is performed for the composite landing gear with identical material properties and configurations. The finite element method simulation is performed on ABAQUS CAE, in which appropriate material properties of cross-weaved Carbon fibre of 400 gsm is stored, followed by layer selection, input of fibre orientation data and finally load application [6].

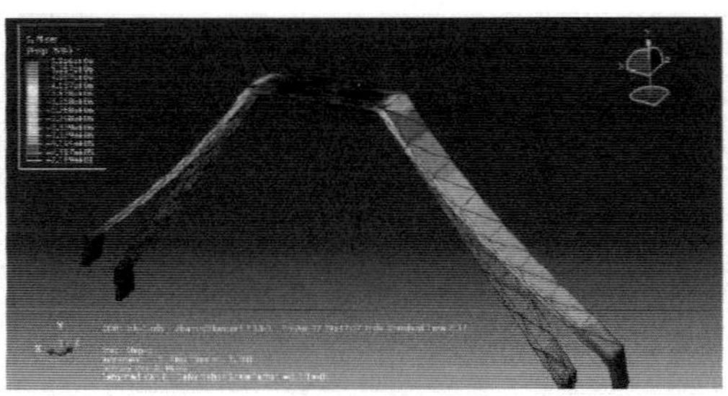

Figure 7. Strain energy contours on landing gear

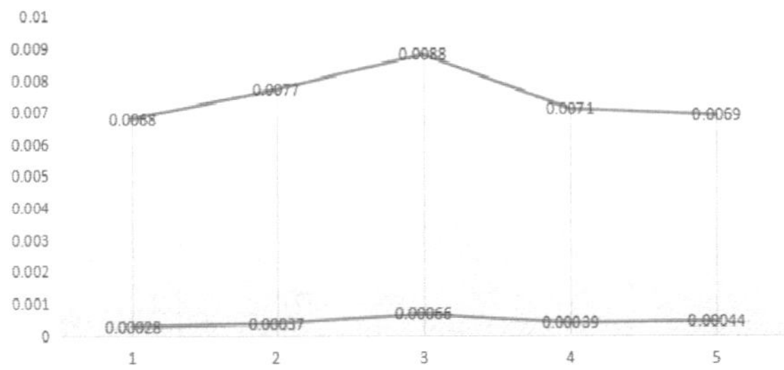

Figure 8. Variations in strains & curvature (FEM)

The results obtained from the finite element model are found to be in close proximity of the results obtained from the mathematical model.

5. Results

The results tabulated below are obtained through the mathematical model developed in MATLAB which represents the mid-plane strains and mid-plane curvatures for each of the 5 composite layups. Layup 1 [0/90/0/90] was found to be the optimal orientation of the fibres in the composite.

Table 2. Mid-plane strains and curvatures for the 5 layups through the mathematical model

Layup	e_x	e_y	e_{xy}	Kx [m]	Ky [m]	Kxy [m]
1	0.000112	0.000188	0.000295	-0.003096	-0.006721	-0.00715
2	0.000223	0.000199	0.000355	-0.004112	-0.005133	-0.00777
3	0.000235	0.000201	0.000344	-0.004215	-0.007541	-0.00812
4	0.000209	0.0002	0.000327	-0.004311	-0.007001	-0.00716
5	0.000156	0.000211	0.000367	-0.003577	-0.00723	0.00738

The results obtained by performing the finite element simulation on the composite landing gear to calculate the mechanical variables are shown in Table 3.

Table 3. Mechanical variables from the finite element simulation

Layup	Max principal stress [Mpa]	Max Strain Energy [J]	Max Equivalent strain	Mean curvature [m]
1	85.7	8.22e9	0.00028	0.0068
2	89.5	8.81e9	0.00067	0.0087
3	80.7	8.34e9	0.00034	0.0077
4	91.7	8.93e9	0.00073	0.0091
5	89.7	8.82e9	0.0070	0.0088

The absolute mean errors obtained for the values of mid-plane strains and curvatures from the 2 methods in Layup 1-5 are approximately 5%, 9%, 7.5%, 12% and 11% respectively.

6. Fabrication

The laminates are fabricated using Vacuum Bagging technique [7] in which the composite material layers are stacked on top of each other using the matrix (Epoxy with a 1:10 ratio of resin and hardener) and then vacuum is created using a sealed plastic bag connected to a vacuum pump. The temperature during the process needs to be increased to accelerate the curing time of the whole layup. This technique is dependent on season temperature, wherein the curing time varies by a large margin in summers compared to winters [7].

Figure 9. Composite layup using carbon fibre

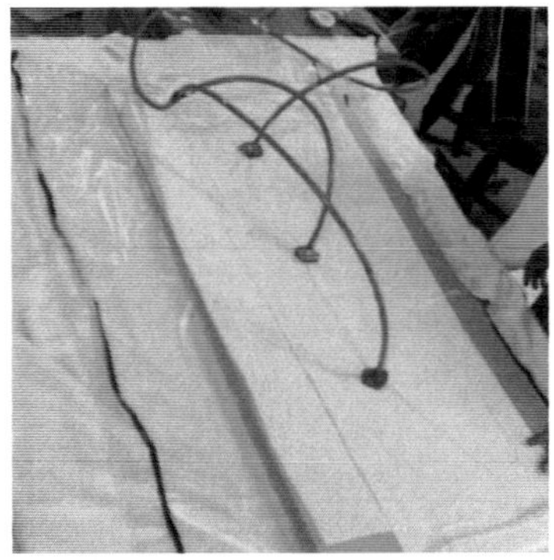

Figure 10. Sealed vacuum layup

The laminates fabricated are then used for physical testing through a 3 point flexural test, before fabricating the actual model. The landing gear, initially designed and followed for a static load of 85N and a dynamic load of 150N behaved structurally well beyond the specified limits, giving a margin of safety of 1.7.

7. Conclusion

The conducted research presents an outline for the engineering analysis of composites in a UAV landing gear. The results of the formulated mathematical model are in close agreement with the Finite Element simulation performed on ABAQUS CAE for 5 different composite layups considered for the landing gear. Fabrication of the small scale laminates for 5 layups followed by a 3 point flexural test validates the design of the composite landing gear. Vacuum Bagging technique has been utilized in the fabrication process of the laminates as well as the actual model. Finally, for a UAV weight of 7.7 kilograms, the landing gear was fabricated with a total weight of 400 grams and reaching a strength/weight percentage of 160%. The technical readiness level of the developed system is observed to be 5.

References

[1] Sadraey M., *Aircraft Design: A Systems Engineering Approach, 2012, Wiley Publications*
[2] Autar K. Kaw: *Mechanics of composite materials, 2nd Edition,*2006 by Taylor & Francis Group, LLC
[3] https://www.asminternational.org/documents/10192/1849770/05287G_Sample_Chapter.pdf
[4] https://snebulos.mit.edu/projects/reference/MIL-STD/MIL-HDBK-17-2F.pdf
[5] Chen, C.H. and Cheng, S. (1967). *Mechanical Properties of Fibre Reinforced Composites,* Journal of Composite Material, 1(1): 30–40.
[6] Klaus Jurgen Bathe: *Finite Element Procedures, 1996 by Prentice Hall, Inc.*
[7]https://www.amtcomposites.co.za/sites/default/files/media/howto/Intro%20to%20Vacuum%20Bagging.pdf

IOP Conf. Series: Materials Science and Engineering **422** (2018) 012006 doi:10.1088/1757-899X/422/1/012006

Laminated CFRP Cutting Area Deformation Analysis and Sub-surface Damage Control

MENG Qing Xun[1, a], JIANG Shou Shan[1,b]

[1]127 West Youyi Rd 704 Hangkong building, NWPU, Xi'an, Shaanxi, China 710072

E-mail: [a]qingxun_m@163.com, [b]shoushanjiang@gmail.com

Abstract: Carbon fiber-reinforced polymer (CFRP) has been used in many aspects. During machining of CFRP damages are always brought in. Due to the brittleness property, the damages are closely related to the deformation in cutting area. In this study, a three-dimensional finite element model was established to study the deformation transformation in cutting area and the interactions of the deformation between two adjacent plies. An experiment was also conducted to measure cutting area deformation and sub-surface damage. The simulation results and experimental data were combined to analyze the correlation between cutting area deformation and sub-surface damage. The sub-surface damage was found have a positive relation with cutting area deformation. Based on the analyzing, a stacking strategy was proposed to reduce the damages in sub-surface. Then the optimization was verified in both numerical and experimental way.

1. Introduction

Carbon Fiber-reinforced Polymers (CFRP) has gained wide application in aero and aviation industries. The high durability and long service life of products require the machining of CFRP parts with extraordinary high precision and least defects. During mechanical machining, however, though the dimensional precision may meet design requirements, when examining the sub-surface burr, delamination and fiber-matrix interface debonding occur frequently [1]. The micro-scale damages accumulate and propagate to form the macro-scale cutting area deformation. The cutting area refers to the area that deforms under the cutting load applied by cutting edge. Due to the brittleness property of CFRP, even a tiny deformation may lead to irreversible damages in fibers and matrix. Therefore, investigating the cutting area deformation is of great importance to reveal the formation and propagation of damages, and to propose methods for minimizing the damages.

In literature, the relation between damages and cutting area deformation is not well explained. Ref. [2-6] established FEM models to simulate CFRP cutting responses. Equivalent homogeneous material (EHM) was used to characterize CFRP and Hashin criteria was used to describe failure initiation. The models were used to study sub-surface damage [2], delamination [3], chip formation [4-5] and cracks propagation [6]. The deformation in the whole cutting area are not paid enough attention.

When refers to cutting area deformation measurement, digital image correlation (DIC) method is commonly used. The exterior surface deformation in cutting area can be measured using DIC method. The interior deformation or damage, however, are not able to measure using DIC method. In this study, a novel experiment was conducted to specify the interior sub-surface damage. Based on the numerical and experimental results, a damage control strategy is present.

2. Cutting area deformation simulation

In order to simulate the propagation and interaction of the deformation in inter-plies, an FEM model was built using Abaqus/Explicit according to ref. [3]. The properties of EHM were offered by the manufacturer (Weihai Guangweicompanies Co.Ltd) as shown in Table 1. The Hou criteria presented in ref. [3] was implemented by VUMAT to describe materials failure. Two plies were extracted to run the simulation. Figure 1 shows the cutting regime and boundary condition of the FEM model. The two plies were tied together. Fiber orientation angle (FOA) was defined as the angle between cutting direction and fiber direction, as shown in Figure 1. By changing the principal direction in each ply, the cutting area deformation for any FOA can be simulated. When investigating the interaction of adjacent plies, the FOA of the two plies were set 45° and 135° respectively. The rake and clearance angle of the cutting tool were both 10°. The cutting depth was 100μm. The cutting tool was selected as analytical rigid body and a reference point was set on it. The velocity load was acted on the reference point to simulate real cutting speed. The mesh of local cutting area was refined to increase computing accuracy. The element type was 8-node brick elements with reduced integration (C3D8R). The element size was 4μm in refined zone. The interaction between rake face and elements was surface-to-node contact. The friction coefficients of the two surfaces were 0.3, 0.8, 0.49 and 0.6 respectively when FOA=0°, 45°, 90°and 135°.

Table 1. Properties of EHM

Properties	value
Elastic modulus (Gpa)	E_{11}=120, E_{22}=8
Poisson's ratio	v=0.25
Longitudinal/transverse tensile strength (Mpa)	X_T=1800/Y_T=55
Longitudinal/transverse compressive strength (Mpa)	X_C=1200/Y_C=200
Shear strength (Mpa)	S_L=100/S_T=80

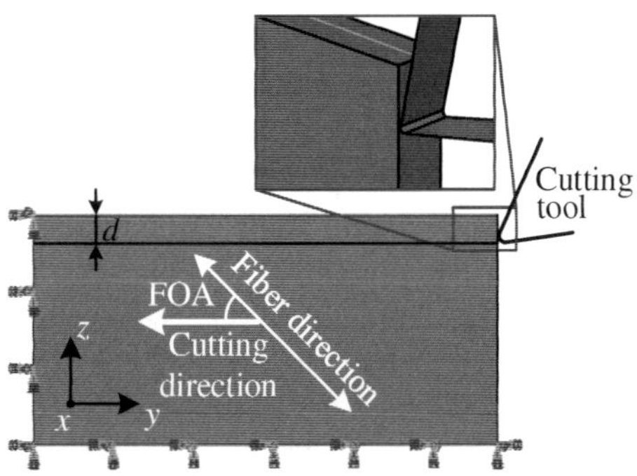

Figure 1. The scheme of FEM model

3. CFRP cutting experiment

CFRP cutting experiment was conducted to measure cutting area deformation and the damage propagation in sub-surface. The CFRP used in experiment was T300/7901. The thickness of each ply was 0.2mm with 56.3% fiber volume fraction. Two types (T_u and T_m) of workpieces were made. The stacking sequence of the T_u type was $[0°]_{20}$, while the stacking sequence of the T_m type was $[45°_{10}/\text{-}45°_{10}]$. Both the dimension of T_u and T_m were 60×60×4mm. By aligning the workpieces in different direction, the FOA of T_u type can be 0°, 45°, 90° or 135°, and the FOA of T_m type was a combination of 45° and 135°. T_u type workpieces were used to measure the cutting area deformation when only one FOA was involved, while T_m type workpiece was used to investigate the deformation of the orthogonal stacking plies with two different FOAs. For T_u cutting, digital image correlation (DIC) method was used to measure the deformation. Random spots were sprayed on the surface which facing to the camera, as shown in Figure 2. The deformation was measured by calculating the movement of the spots [7]. For T_m cutting, the sub-surface in the adjacent area of the two plies was examined to qualify the transverse transfer of deformation. A small cube which contained the adjacent area was taken out for the purpose of inspection.

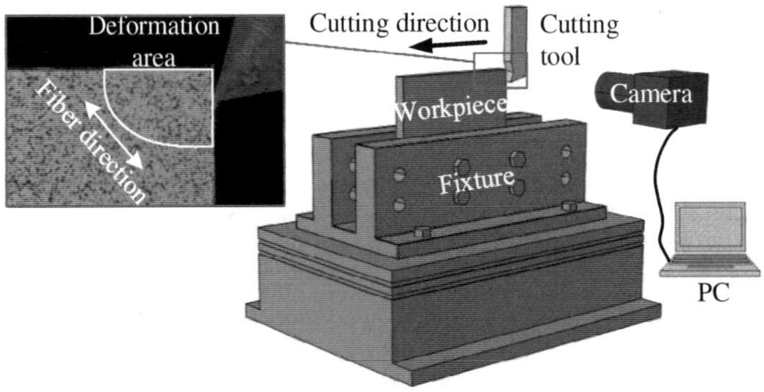

Figure 2. Scheme of experimental setup

4. Results and discussion

Figure 3 shows the comparison of the numerical and experimental results. The experimental results were obtained using T_u type workpiece. A deformation area forms in front of the cutting tip as the tool contacts the workpiece. The propagation of the deformation varies according to FOA and its trend is observed propagating along fibers direction in both numerical and experimental results. When FOA=0°, as shown in Figure 3 (a) and (e), the deformation occurs not only in the materials to be removed but also in sub-surface. According to the DIC results in Figure 3 (e), the cracks propagate exactly in front of the cutting tip. The deformation, however, occurs not only in the materials to be removed, but also transfers to the sub-surface. When examining the sub-surface, as shown in Figure 3, no damage was observed.

As FOA increases to 45°, the deformation in cutting area was found transferring along fibers direction. This can be seen from both numerical and experimental results, as shown in Figure 3 (b) and (f). The largest deformation occurred in front of the cutting tip and then it decreased gradually in fibers direction. The sub-surface damage was about 20μm (shown in Figure 3 (i)), which is much larger than that in FOA=0°. The cracks were not observed in the experiment since the fibers and matrix around the cutting tip were crushed into powder-like chip.

When FOA reached 90°, more fibers and matrix involved in the deformation. The deformation transferred deeply into sub-surface, as shown in Figure 3 (c) and (g), leading to severe sub-surface damage. This is because the fibers and matrix tended to be bended towards lateral side to avoid the cutting tip. After experiencing out-of-plane deformation, extensive fiber-matrix interface debonding occurred in the cutting area. The damage length in sub-surface reached 235μm in this situation (shown in Figure 3 (k)).

When FOA is 135°, the fibers were raised to slide on the rake face as contacting the cutting tool. The role of cutting tip was weaken to a large extent. Larger out-of-plane deformation and severer fiber-matrix debonding were observed. The deformation in cutting area transferred deeply in the sub-surface, as shown in Figure 3 (d) and (h). The damage length in sub-surface reached 590μm, which is the largest in the four situations.

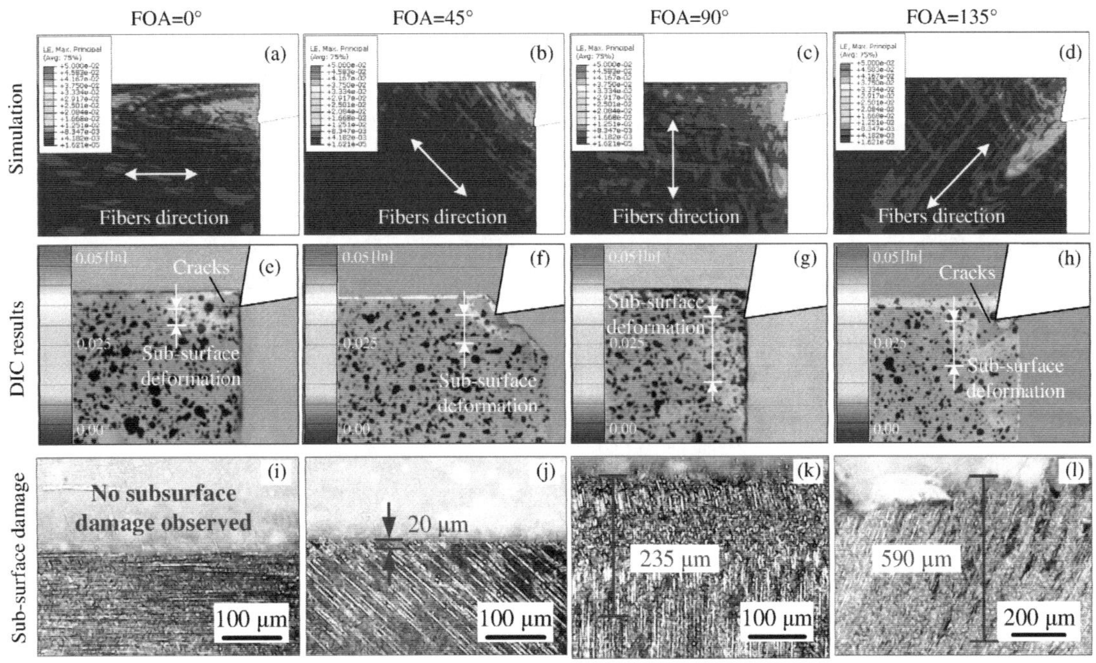

Figure 3. Comparison of cutting numerical and experimental results

According to the simulation and experimental results, the sub-surface damage has a positive correlation with the deformation in cutting area. One potential way to diminish sub-surface damage is to control the deformation in cutting area. For this purpose, an orthogonal stacking strategy was proposed in this study. Figure 4 shows the stress distribution in cutting area when the fibers in two adjacent plies are parallel and perpendicular. As can be seen, the stress in orthogonal stacking plies is not as concentrate as that in non-orthogonal stacking plies. This means the deformation in cutting area is reduced after the interaction of the two orthogonal stacking plies. In order to examine the sub-surface damage, a cutting experiment was conducted using T_m type workpiece. A small cube was taken from the workpiece. Surface 1-2-3-4 was the machined surface, while surface 3-4-5-6 was a cross-section in the cutting area. After polishing the cube, it can be seen an obvious boundary between the two orthogonal stacking plies. The damages cannot be judge from the machined surface. When examining surface 3-4-5-6, however, it was found that there was almost no damage for FOA=45° but a severe damage for FOA=135°. The damage reduced gradually from the middle to the right side, as shown in Figure 5. This means the deformation can be controlled by adjusting stacking sequence since the deformation in two adjacent plies interacted with each other.

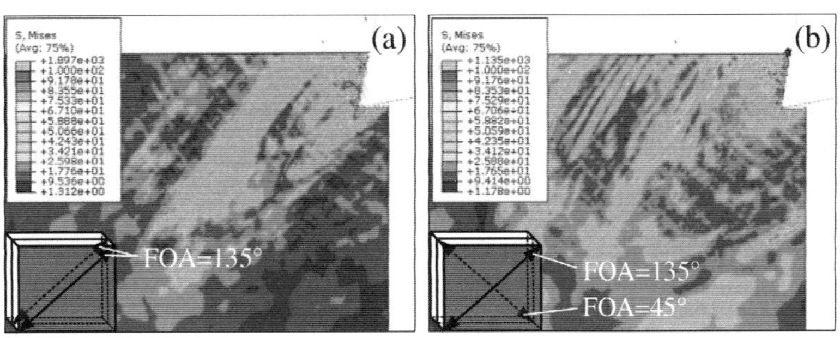

Figure 4. The stress distribution in cutting area: (a) non-orthogonal stacking; (b) orthogonal stacking

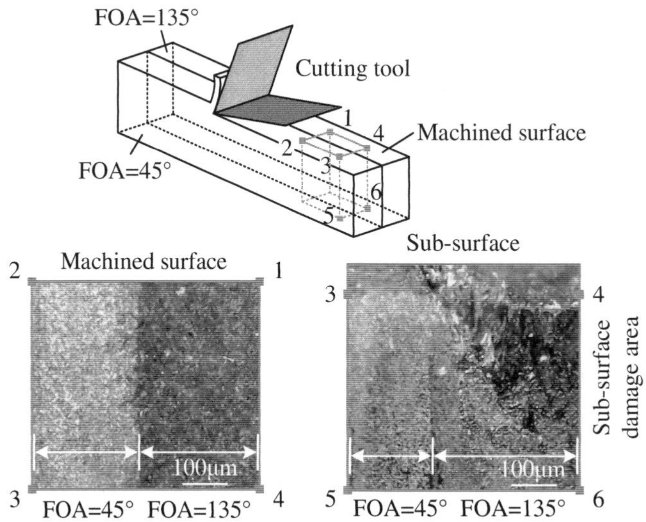

Figure 5. Machined surface and sub-surface damage

5. Conclusions

A numerical CFRP cutting model was built to simulate the stress disribution and an experiment was conducted to measure deformaiton in cutting area. According to numerical and experimental results, it can be concluded that the deformation in cutting area transfers along the fibers direction. The sub-surface damage has a positive correlation with the deformation. The interaction between two orthogonal adjacent plies can be used to control sub-surface damge. By applying the strategy in this study, the damage during machining of CFRP will be reduced to a large extent.

6. References

[1] M Mohit, K Mahesh, S Ranjeet. An approach towards machining of GFRP using alumina based cutting tools. International Journal of Materials Science and Engineering. 2016, 4(4): 185-191.

[2] J Yin, Z Jia, F Wang, et al. FEM simulation analysis of subsurface damage formation based on continuously cutting process of CFRP. Journal of mechanical engineering. 2016, 52(17): 58-64 (in Chinese).

[3] C Santiuste, H Miguélez, X Soldani. Out-of-plane failure mechanisms in LFRP composite cutting. Composite Structures. 2011, 93(11): 2706-2713.

[4] S Zenia, A L Ben, M Nouari, et al. Numerical analysis of the interaction between the cutting forces, induced cutting damage, and machining parameters of CFRP composites. International Journal of Advanced Manufacturing Technology. 2015, 78(1-4): 465-480.

[5] J Xu, ME Mansori. Numerical modeling of stacked composite CFRP/Ti machining under different cutting sequence strategies. International Journal of Precision Engineering and Manufacturing. 2016, 17(1): 99-107.

[6] S Usui, J Wadell, T Marusich. Finite Element Modeling of Carbon Fiber Composite Orthogonal Cutting and Drilling. Procedia CIRP, Dornfeld D, Helu M, 2014: 14, 211-216.

[7] LIU X Y. Digital image correlation method and its application on mechanical properties measurement of materials. Chang chun: Jilin university, 2012.

IOP Conf. Series: Materials Science and Engineering **422** (2018) 012007 doi:10.1088/1757-899X/422/1/012007

Damage Inspection and Online Monitoring using Lamb Waves: A Comparative Study on Aluminium and Composite Plate Structures

Royson Dsouza [1,a], Anil Sequeira [1,b], Marian Jose [2] and Gautami Golani [2]

[1] School of Engineering and IT, Manipal University Dubai, Dubai, UAE.
[2] Mechatronics Department, School of Engineering and IT, Manipal University Dubai, Dubai, UAE.

E-mail: [a] roysonnitk@gmail.com, [b] anil.sequeira@manipaldubai.com

Abstract: Structural damage is an important criterion that can affect the system performance and degrade its ability from its original condition. When it comes to an aircraft, it is the wings of the aircraft that is prone to more damages compared to the body since the wings provide overall structure stability to the aircraft. It is thus mandatory to monitor the surface of the aircraft wing periodically in search of new damages or changes occurring in the wing structures. The term Health Monitoring systems is introduced in these precepts, that are useful in identifying damages in aircrafts, space crafts and mechanical infrastructures by taking periodic measurements with the help of various transmitters and sensors. Waves are allowed to propagate through the wing surface and the deviations occurring in the frequency of the waves are recorded and compared which helps in determining the various types of damages that occur. Yet another consideration is that the sensors embedded to the aircraft wing must be light weight, cost effective, user friendly at the same time provide accurate results for easy diagnosis of the problem. This is achieved by making use of lamb waves (also named as guided plate waves). The proposed work aims at the health monitoring of aerospace and mechanical structures, by using the wave propagation method and the use of piezo-electric wafer active sensors. Results have been validated by experimental and analytical calculations and the proposed method shows enhanced performance in the detection of cracks on the structural surfaces.

1. Introduction

Health monitoring signifies a system which can determine undesirable conditions or damages in the structure so as to improve its durability and life-cycle, at the same time reducing the overall maintenance cost. The aerospace industry uses this health monitoring system as regular inspections and repair of aircrafts can prove to be very expensive. To reduce costs, composite materials are being increasingly used as they have excellent specific strength and specific modulus characteristics. Along with this, they are lightweight, resistant to corrosion and chemical attack and also offer better performance than most metals at high temperatures. Damage detection in composite laminates is difficult due to its anisotropy. Non-destructive evaluation (NTE) methods are used for assessing the safety of the composite. One such method is C-scan, but it is unrealistic for the examination of large components. Others like Eddy Currents, Visual inspection, X-ray radiography and Ultrasounds are expensive, time-consuming and can be performed on surfaces that are accessible only.

Lamb wave based structural health monitoring is widely used in recent times. Lamb waves are guided acoustic waves that propagate in the plane of the structure, which according to its displacement patterns, it can be divided into symmetric Sn and anti-symmetric An lamb waves with respect to the mid-plane of the structure [1]. Lamb wave find its potential application for a large area Non-Destructive Investigation (NDI) due to the advantage of Lamb wave being able to transmit and receive for long distances.

Some of the universally used transducers for non-destructive testing are angle piezoelectric wedge transducers [2] [3], comb transducers and electromagnetic acoustic transducers (EMATs) [4]. Also used are Hertzian contact transducers and lasers. These transducers are efficient for performing maintenance checks when the structure is offline for servicing, but not compact enough to be permanently mounted on-board the structure. Surface-bonded or embedded piezoelectric wafer transducer is one of the very universally used transducers for structural health monitoring. Piezoelectric transducers are inexpensive and are manufactured in sizes of incredibly low thickness, and can thus be easily integrated into structures without causing much obstruction or inconvenience. They further couple the electrical and automated behaviour of the material being monitored. The piezoelectric effect and its inverse effect act as the principles that manage their operation. The major materials universally offered are lead zirconium titanate ceramic, also known as PZT and a polymer film namely, polyvinylidene fluoride (PVDF), designed by Monkhouse [5] [6]. Therefore, PZT is the very commonly used for health monitoring [7] [8] [9] [10].

An acoustic based structural health monitoring (SHM) system was developed for composite structures [11]. The construction of Low Profile Acoustic Transducer (LPAS) is similar to angled wedge ultrasonic transducers that are used for offline NDT. These are used to generate Guided Waves (GWs). The size of these transducers are much smaller when compared with conventional ultrasonic transducers, and thus they possess higher possibility of using them as on-board SHM transducers.

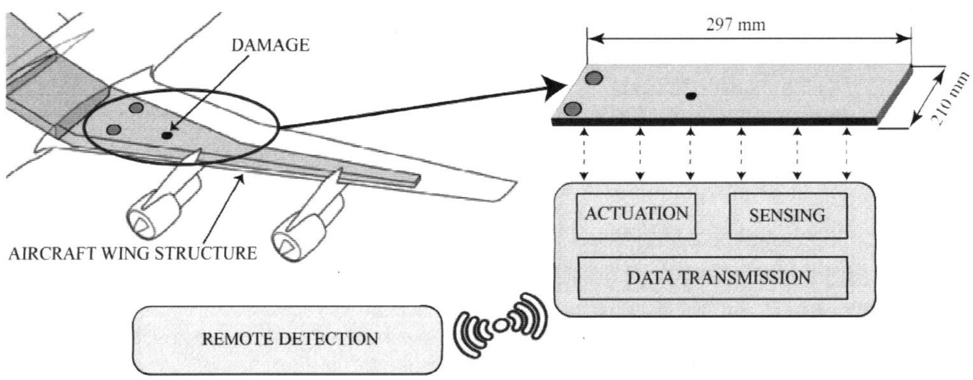

Figure 1. Concept of Structural Health Monitoring using Lamb Waves

The use of guided waves in structural health monitoring and its effects of elevated temperature in internal spacecraft structures is studied and presented [12]. A detailed review on the use of non-linear guided wave in the structural health monitoring on real-time implementation of industry problems is carefully studied [13]. The authors in [14] have shown the creation of damage index for evaluating the extent of damage in composite laminates using guided waves. The debonding effect is more precisely investigated with mathematical simulations and experimental models on the piezo-induced wavefield [15]. The concept of structural health monitoring using lamb waves is shown in Figure1. A detailed study on actuation and sensing properties of Piezoelectric surface mounted transducers at various temperatures is carried out for Lamb wave based damage detection. Experimental validation is also provided for the same [16].

The primary objective is to fabricate Carbon Fiber Reinforced Plastics (CFRP) plates (297×210mm) of 4-ply and propagate lamb waves through these plates with and without crack and later interface the signal that is output from the digital oscilloscope to a laptop with LabView installed in it to detect the

position of the crack. ABAQUS 6.14 is used for simulation of the lamb waves through the CFRP plates in order to match the ABAQUS results with the experimental results and later identify the percentage error or the deviation occurring of experimentally generated results from the theoretical.
The Aluminium plate (thickness 1.2mm) tested results are also compared with the CFRP plates in order to study the wave nature of both the materials. The experiment was also performed on Aluminium and CFRP plates of varied plys by subjecting linear wave scans on them. The transmitted and reflected waves were monitored with circular disk type piezoelectric sensors. ABAQUS models have been developed to select actuator configurations that are ideal for the actuating signals generated in the experiments.
The experimental and theoretical results have been compared among the models developed and they have been used to identify the percentage error that helps in predicting the damage detection capabilities. The techniques employed using Lamb wave provide localized information about the structural and are thus well suited for health monitoring applications.

2. FEA Modeling and Analysis

A 3D Finite element modelling and analysis was carried out using Abaqus FEA software to determine and analyze the response of Aluminium and CFRP specimens subjected to damage and undamaged conditions. Dynamic explicit module being chosen for a time instance of 0.5ms. Approximate global size of 0.001 is selected for Global seed mesh size. The excitation direction is provided along the axis of the plane. The FEA is performed for both Aluminium and CFRP specimens.

2.1 Aluminium plates without and with damage

Preliminary analysis is carried out on a Aluminium plate of 1.5mm thickness and dimensions 297mm×210mm attached with a piezo-electric wafer active disk type actuator and a sensor of diameter 7mm. When voltage is applied on to the actuator at a frequency of 150 KHz, lamb waves are generated in omni-direction to the surface and these waves collide with the edges of the plate scatters all over the plate and reaches the sensor which receives these waves.
Lamb waves generated from the piezo-electric actuator propagate through the surface of the plate and spreads all over the plate. When these waves meet the region where damage is present the waves collides and gets reflected as shown in Figure 2. To simulate the damage a hole of 20mm was created exactly at the centre of the plate.

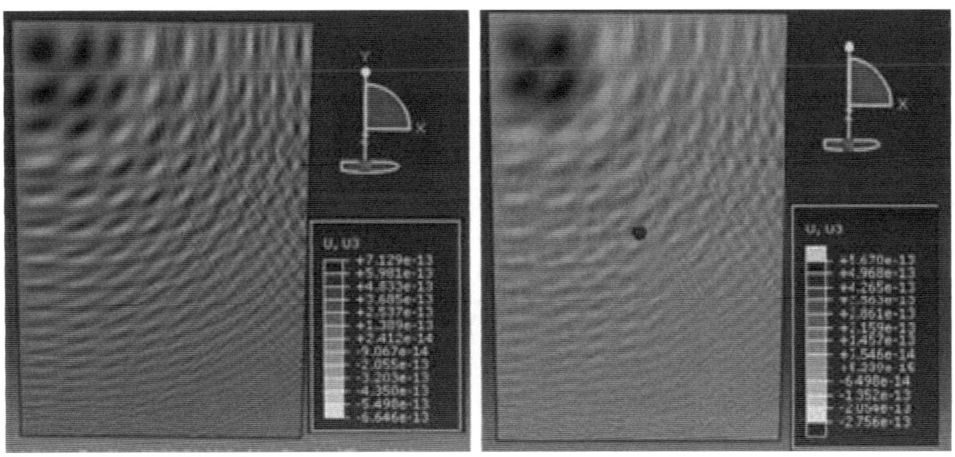

Figure 2. FEA Simulations of Aluminium specimen without and with defect

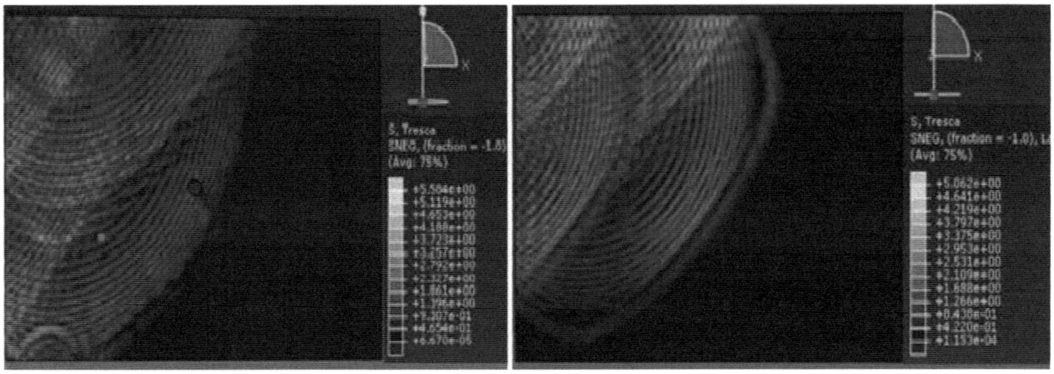

Figure 3. FEA Simulations of CFRP specimen with and without defect

The location of the crack is given by Equation1.

$$\alpha = \frac{L\Delta t_{dam}}{\Delta t_L} \tag{1}$$

Where, Ld= Location of the crack from the left hand edge of the specimen.
L=Length of the specimen.
Δt=Time taken by the reflected lamb wave on the reference specimen.
Δtd Time taken by the reflected wave on the specimen with crack.
From the Figure 4, it is seen that the amplitude peak due to the first reflected wave is seen at a time period of 0.31×10^{-3}s. In graph 4.2, an amplitude variation due to crack is seen at 0.15×10^{-4}s the time taken by the reflected wave. Substituting the parameters the location of the crack is found to be 148mm. Figure 4 clearly shows the amplitude time domain signals of FEA simulation results. The transmitted signal and reflected signal can be clearly distinguished alongside the change of amplitude with respect to time. A change in amplitude is clearly observed when the aluminium plate specimen is subjected to damage. The output response of Aluminium plate specimen without damage is taken as reference signal as shown in Figure 4(a). When the plate is subjected to damage the change in the amplitude is observed in Figure 4(c). Arithmetic difference is the difference between the current signal with the reference signal. Figure 4(d) shows a appreciable arithmetic difference between the signal of Aluminium plate specimen with crack and reference signal.

2.2 Carbon Fibre Plate without and with damage

Carbon fibre reinforced plastics (CFRP) of 4-ply are tested without crack initially and the generated waves travel through the cross-ply and matrix and reach the piezo-electric sensor which outputs the signal to the wave order to compare the results with the crack formation. Lamb waves generated from the piezo-electric actuator propagate through the surface of the CFRP plate and spreads all over the plate. When these waves strike the region where the two cracks are present the waves collide and get reflected as shown in Figure 3. These reflected waves are obtained at the receiving sensor and transferred to the digital oscilloscope which saves the output waveform. Later, the waveforms obtained on the CFRP plate with crack and the plate without crack is compared and the location of the crack is obtained. Substituting the values of variables from the graph in Equation 1 the location of the crack is found to be 147.5mm. Figure 5 clearly shows the amplitude time domain signals of FEA simulation results. The transmitted signal and reflected signal can be clearly distinguished alongside the change of amplitude with respect to time. A change in amplitude is clearly observed when the CFRP plate specimen is subjected to damage. The output response of CFRP plate specimen without damage is taken as reference signal as shown in Figure 5(a). When the plate is subjected to damage the change in the amplitude is observed in Figure 5(c). Arithmetic difference is the difference between the current signal with the reference signal. Figure 5(d) shows a appreciable arithmetic difference between the signal of CFRP plate specimen with crack and reference signal.

3. Experimental Setup

Keeping the input and output requirements, an experimental setup is designed to investigate the behavior of lamb waves in CFRP plate specimen in comparision with Aluminium plate specimen subjected to damage and no damage. The schematic diagram of the experimental setup is shown in Figure 6.The testing for damage has been carried out using Aluminium plates and Carbon Fibre Reinforced Plastics (CFRP). Lamb waves are generated on specimen of size 210×297 mm by applying a sinusoidal wave of 11Vp-p voltage at a frequency of 240KHz to a piezoelectric disc operated in the longitudinal mode by using Scientec 4061 Signal generator. The experimental setup was placed as shown in Figure 7. When voltage is supplied to the piezoelectric disc it expands and contracts parallel to the surface of the plate which produces a bending moment that generates the lamb waves.

The propagation of the generated lamb waves are used for identifying the location of the damage on the basis of time delay caused by the subsequent reflected waves from the specimens boundaries. Tests were performed on specimens made of Aluminium of size 210×297×1.2 mm having density=2.580g/cm^3,shear modulus=23GPa, youngs modulus=72GPa. The plate was instrumented using two piezoelectric discs of size 20 mm diameter made of commercial brass backed piezo ceramic resonators where one PZT was used as an actuator (PZT 1) and the other used as a sensor (PZT 2) bonded at the end of the plate as shown in Figure 6.

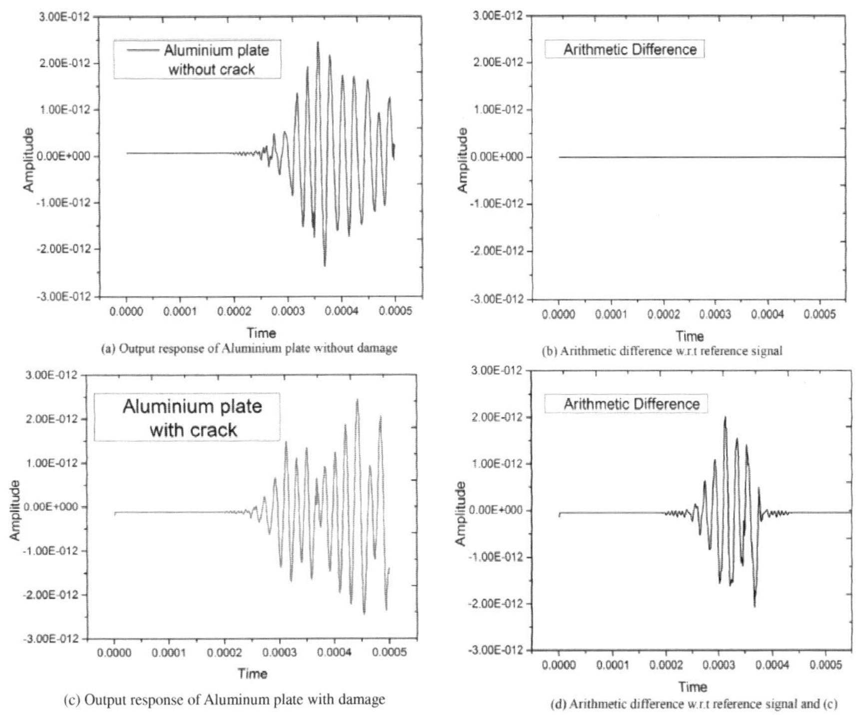

(a) Output response of Aluminium plate without damage
(b) Arithmetic difference w.r.t reference signal
(c) Output response of Aluminum plate with damage
(d) Arithmetic difference w.r.t reference signal and (c)

Figure 4. FEA Simulated response Arithmetic difference between the response signals of the Aluminum specimen

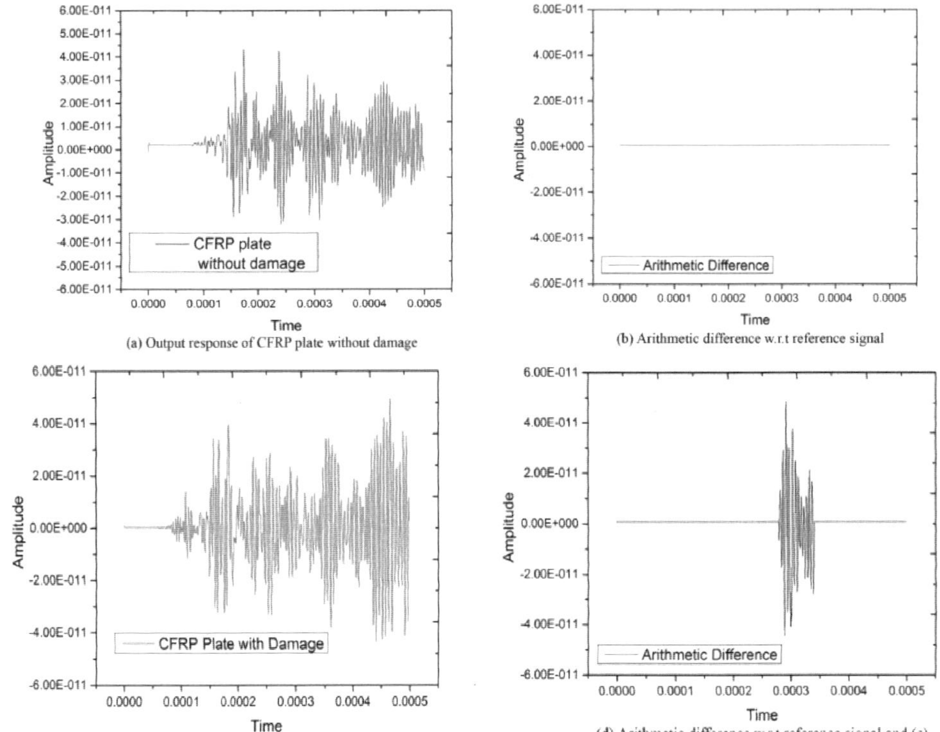

(a) Output response of CFRP plate without damage
(b) Arithmetic difference w.r.t reference signal
(c) Output response of CFRP plate with damage
(d) Arithmetic difference w.r.t reference signal and (c)

Figure 5. FEA Simulated response Arithmetic difference between the response signals of the CFRP specimen

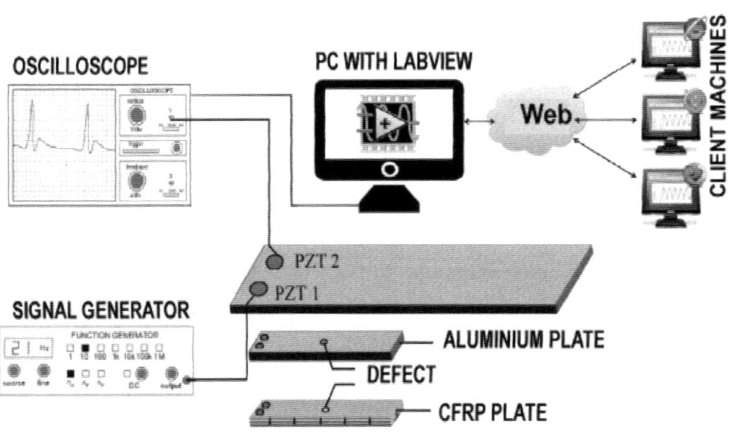

Figure 6. Schematic diagram of the experimental setup

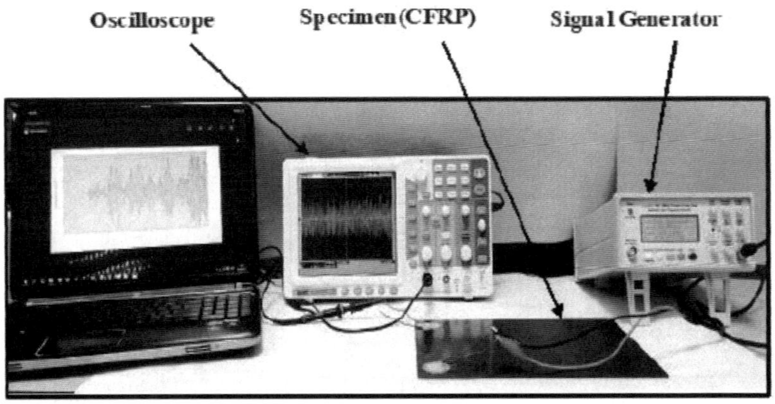

Figure 7. Experimental setup

The same experimental setup and procedure used for the analysis of Aluminium plate is used for the testing of Carbon fibre. All the four test plates were tested by following the aforementioned procedures and the waveform obtained by the defect-free plate was used as the reference and compared with the waveform obtained by the plate having the defect. The resultant waveform signals were analyzed using LabVIEW. Arithmetic difference of the signals, display of location of crack is developed in front Panel of LabVIEW. The application is then deployed to the webserver which is then remotely monitored using Ethernet Communication.

4. Results and Discussions

Tests are carried on Aluminium specimen and CRFP specimens as mentioned above. The response graphs are recorded in Labview and subsequent graphs are plotted for the same and are discussed in following sections.

4.1 Experimental Result Analysis of Aluminium specimen

Figure 8 shows the response signals of the tests carried on Aluminium plate subjected to damage and without damage. From the 2 graphs it is very evident that there is a change in response signal from the damaged plate to the undamaged plate at a given time interval. Location of the damage can be found using the Equation 1. From Figure 8 it is very evident that the amplitude peak due to the first reflected wave is seen at a time period of 0.33×10-4s and amplitude variation due to crack is seen at 0.16×10-4s the time taken by the reflected wave.Using Equation 1 the location of the graph is found to be 144mm. Figure 9 clearly indicates the arithmetic mean difference of the Aluminium specimen with and without defect.

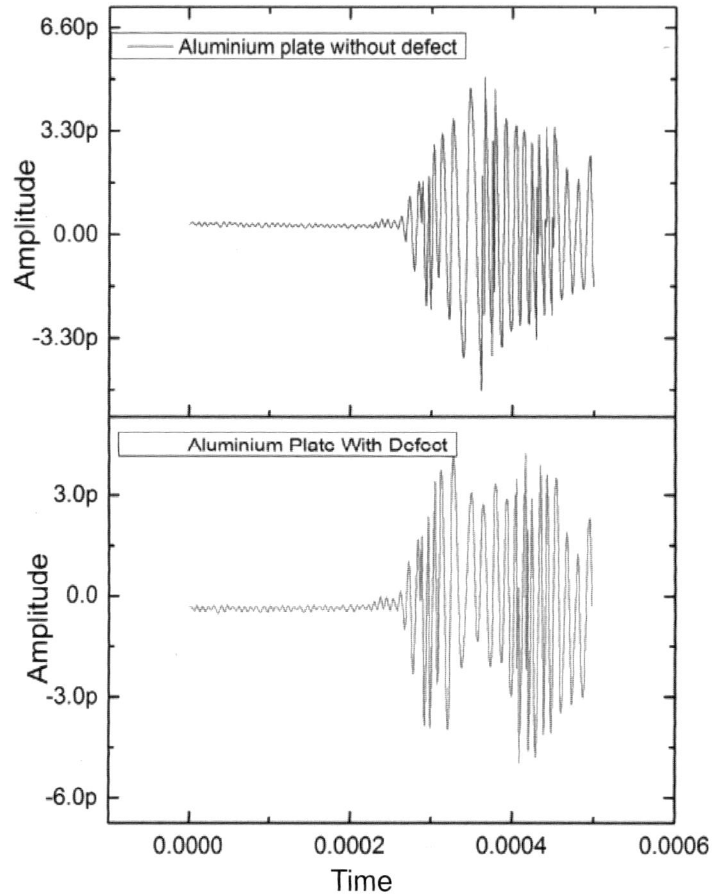

Figure 8. Response signals for the Aluminium plate with and without defect

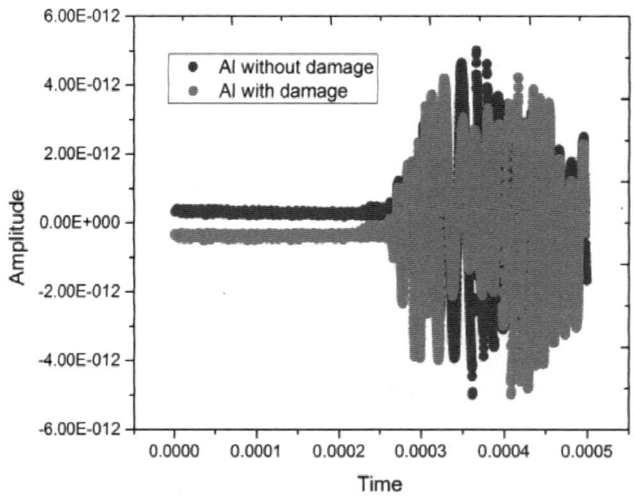

Figure 9. Comparison of response signals from Aluminium plate with and without defect

4.2 Result Analysis of CFRP specimen

From the graph shown in Figure 10, it is seen that the amplitude peak due to the first reflected wave is seen at a time period of $0.25\times10\text{-}4^{s}$ and an amplitude variation due to crack is seen at $0.12\times10\text{-}4^{s}$ the time taken by the reflected wave. From Equation 1 the location of the crack is calculated to be 142.56mm.

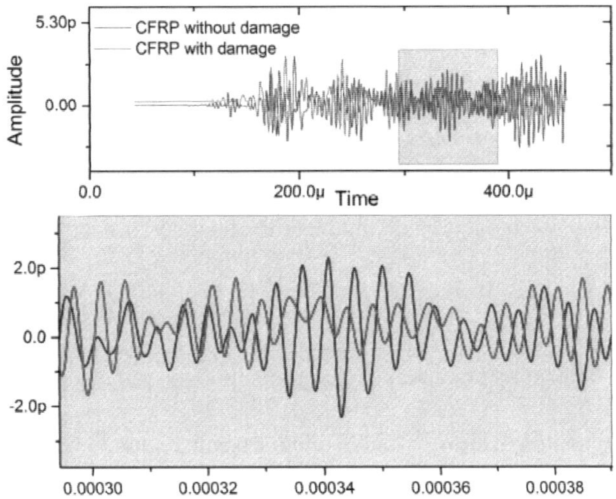

Figure 10. Response signals for undamaged and damaged CFRP specimen

5. Conclusion and Future Work

Health monitoring of composite laminates using piezoelectric wafer active sensors was proven a reliable, robust, light weight and low budget method to generate the lamb waves compared to other costly methods such as ultrasonic methods which require large comb or wedge transducers to carry out the scanning across the plate.

Lamb wave methods have good potential for implementing in health monitoring system. This method provides useful information about the presence and extent of damage in both aluminium and composite materials at the same time hold the capability of determining the location and type of damage which helps to fixed onto a structure with conformable piezoelectric devices.

The major disadvantage of lamb waves is that it is active, it requires voltage supply and function generating signal to be supplied the signal. This can be complicated in a large structure, especially if the health monitoring system is to be implemented wirelessly. Remotely activated PZT using radio frequency waves can be used for future work in order to implement this type of method in a large structure.

6. References

[1] Z. Su and L. Ye, *Identification of damage using Lamb waves: from fundamentals to applications*. Springer Science & Business Media, 2009, vol. 48.

[2] W. Zhu, J. Rose, J. Barshinger, and V. Agarwala, "Ultrasonic guided wave ndt for hidden corrosion detection," *Journal of Research in Nondestructive Evaluation*, vol. 10, no. 4, pp. 205–225, 1998.

[3] P. Wilcox, M. Lowe, and P. Cawley, "Mode and transducer selection for long range lamb wave inspection," *Journal of intelligent material systems and structures*, vol. 12, no. 8, pp. 553–565, 2001.

[4] G. Alers and L. Burns, "Emat designs for special applications," *Materials Evaluation*, vol. 45, no. 10, p. 1184, 1987.

[5] R. Monkhouse, P. Wilcox, and P. Cawley, "Flexible interdigital pvdf transducers for the generation of lamb waves in structures," *Ultrasonics*, vol. 35, no. 7, pp. 489–498, 1997.

[6] R. Monkhouse, P. Wilcox, M. Lowe, R. Dalton, and P. Cawley, "The rapid monitoring of structures using interdigital lamb wave transducers," *Smart Materials and Structures*, vol. 9,

no. 3, p. 304, 2000.

[7] C.-T. Ng and M. Veidt, "A lamb-wave-based technique for damage detection in composite laminates," *Smart materials and structures*, vol. 18, no. 7, p. 074006, 2009.

[8] A. Purekar and D. Pines, "Damage detection in thin composite laminates using piezoelectric phased sensor arrays and guided lamb wave interrogation," *Journal of Intelligent Material Systems and Structures*, vol. 21, no. 10, pp. 995–1010, 2010.

[9] J. Moll, M. V. Golub, E. Glushkov, N. Glushkova, and C.-P. Fritzen, "Non-axisymmetric lamb wave excitation by piezoelectric wafer active sensors," *Sensors and Actuators A: Physical*, vol. 174, pp. 173–180, 2012.

[10] H. Boukabache, C. Escriba, and J.-Y. Fourniols, "Toward smart aerospace structures: Design of a piezoelectric sensor and its analog interface for flaw detection," *Sensors*, vol. 14, no. 11, pp. 20 543–20 561, 2014.

[11] S. G. Pierce, F. Dong, K. Atherton, B. Culshaw, K. Worden, G. Manson, T. Monnier, P. Guy, J.-C. Baboux, J. Assad et al., "Damage assessment in smart composite structures: the damascos programme," *Air & Space Europe*, vol. 3, no. 3-4, pp. 132–138, 2001.

[12] A. Raghavan, "Guided-wave structural health monitoring," Ph.D. dissertation, The University of Michigan, 2007.

[13] M. Mitra and S. Gopalakrishnan, "Guided wave based structural health monitoring: A review," *Smart Materials and Structures*, vol. 25, no. 5, p. 053001, 2016.

[14] S. Torkamani, S. Roy, M. E. Barkey, E. Sazonov, S. Burkett, and S. Kotru, "A novel damage index for damage identification using guided waves with application in laminated composites," *Smart Materials and Structures*, vol. 23, no. 9, p. 095015, 2014.

[15] M. Golub, A. Shpak, I. M¨uller, and C.-P. Fritzen, "Numerical simulation of lamb wave excitation by the partially debonded rectangular strip-like piezoelectric actuator based on the integral approach and hp-fem," in *Days on Diffraction* (DD), 2016. IEEE, 2016, pp. 171–176.

[16] P. Kijanka, P. Packo, X. Zhu, W. J. Staszewski, and F. L. di Scalea, "Three-dimensional temperature effect modelling of piezoceramic transducers used for lamb wave based damage detection," *Smart Materials and Structures*, vol. 24, no. 6, p. 065005, 2015.

IOP Conf. Series: Materials Science and Engineering **422** (2018) 012008 doi:10.1088/1757-899X/422/1/012008

Calculation and Analysis of Graphene Nanoribbons Field Effect Transistor Structure and Characteristics

Wei Chun[1], Li Ming[2] and Hang Tongnan[1]

[1]School of Electronic and Information Engineering,Wuhan Donghu University, Wuhan 430211, China
[2]Information management center, Air Force Early Warning Academy,Wuhan 430019,China

Abstract. Because there is no gap between conduction band and valence band of graphene, it is difficult to achieve switching characteristics when transistors are fabricated. In order to open the graphene band gap, researchers have explored many methods, such as tailoring Shi Mocheng quantum dots, nanobelts, nano grids or laying graphene on a special substrate, one of the feasible ways is to regulate the electrical properties of graphene by doping. Through the analysis on the structure and properties of graphene band calculation, with the help of computer aided simulation, the ID and VG characteristics analysis of the Schottky barrier graphene field effect transistor (SBFET) and MOS structure of graphene field effect transistor (MOSFET) are investigated in this study. The results are of high reference value for the study of the structure and properties of graphene nanoribbons.

1. Introduction
Graphene is a two-dimensional film with only one carbon atom thickness, which is composed of carbon atoms with hybridized orbits and six angle honeycomb lattices. The graphene material is very thin, but the hardness is very high, very good transparency is very low, the light absorption, good thermal conductivity, the thermal conductivity than diamond and carbon nanotube; the graphene material has high mobility and low resistivity, mobility than carbon nanotubes and silicon the high resistivity of copper and silver materials is lower, the minimum resistivity material; the resistivity of graphene is low, the conduction speed is very fast, so it can be the development of ultra-thin electronic components and electric liquid crystal materials, has a very good application prospect.

2. Graphene and its application
With the continuous rise of graphene materials, graphene has more extensive applications, including sensors, solar battery materials, lithium ion battery materials, semiconductor and thin film crystal and composite materials, are described in detail in the four main areas of application.

2.1. Sensor applications
The small volume of graphene but a large surface area, high sensitivity and its application in sensor materials, can obtain higher sensitivity than conventional sensors, such as Singapore, Nanyang Technology University developed a graphene material optical sensing sensor, Its sensitivity is 1000 times stronger than that of ordinary graphene sensors.

2.2. Energy storage and new type display

IOP Conf. Series: Materials Science and Engineering **422** (2018) 012008 doi:10.1088/1757-899X/422/1/012008

The electrical conductivity of graphene and the transmission performance is very good, the light can be used in the storage battery and the display screen, touch screen using indium tin oxide materials most of the graphene has very good prospects, it is likely to replace indium tin oxide used in touch screen for future production, Samsung Toshiba and many other well-known mobile phone manufacturers have been locked in this type of material, and focus on the layout of development, I believe that in the near future will develop a new type of touch screen is higher than ordinary screen sensitivity.

2.3. Semiconductor materials

Most of the siliceous material of semiconductor material and semiconductor device at present, strong manufacturing enterprises and research institutions have begun to carry out research and development of graphene, such as the Columbia University has developed a graphene - silicon optoelectronic chip, IBM researchers have developed a graphene field-effect transistor, its performance to manufacturing silicon semiconductor devices compared with the traditional good material.

2.4. Biomedical applications

Graphene nano materials have been applied to the biomedical field. As the carrier of nano drug, biological imaging and detection, it can be used for bacterial analysis, DNA and protein detection, such as the University of Pennsylvania to develop graphene nano hole equipment, greatly improving the detection speed of DNA sequence.

3. Structural properties and energy band calculation of graphene

The electronic band structure of graphene can affect the characteristics of the larger, the whole structure is to separate carbon atoms as the base, the perturbation is generated by the surrounding carbon atoms, the energy level distribution of graphene can be analyzed by matrix calculation, using approximate two quantized Hamiltonian can be expressed in graphite electron tight binding model, such as Formula (1) shown.

$$H = \varepsilon_{2pz}\sum_{i}(a_i^+ a_i + b_i^+ b_i) - t\sum_{<i,j>}(a_i^+ b_j + h.c) \qquad (1)$$

The influence of electron spin is neglected in Formula (1), where $<i,j>$ is the sum of the electron transitions to the neighboring lattice, and ε_{2pz} is the energy of the single electron $2pz$ orbit.

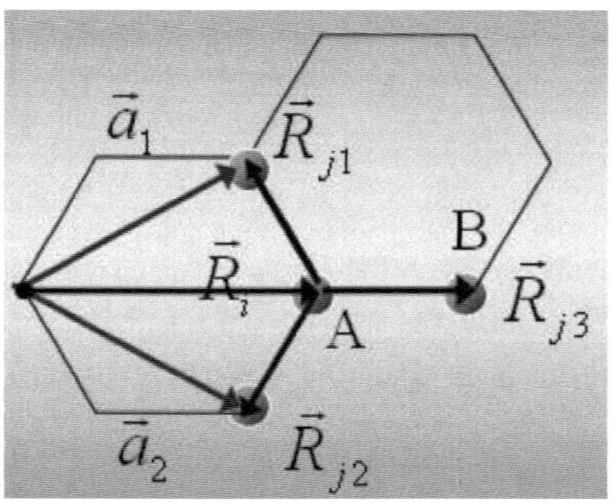

Figure 1. Schematic diagram of adjacent atomic reference atoms

As shown in Figure 1, the graphite lattice can be represented by the two neighboring carbon atoms A, B, the geometrical environment of these two carbon atoms is not equivalent. The A atom is chosen as the reference atom, whose vector is $\vec{R}_i$, and the vectors of the neighboring B atoms are $\vec{R}_{j1}$, $\vec{R}_{j2}$ and $\vec{R}_{j3}$. Suppose that $a_i^+(b_j)$ is generating or destroying operators of electronics at $\vec{R}_i(\vec{R}_j)$. In

Formula (1), this symbol can be expressed as an electron produced by a_i^+ at $\vec{R}_i$ after b_j destroyed an electron at $\vec{R}_j$. It can also be seen as electron transition from $\vec{R}_j$ to nearest neighbor $\vec{R}_i$, the transition energy is $t = 2.8eV$. By Fourier transformation we can get that

$$a_i = \frac{1}{\sqrt{N}} \sum_{\vec{k}} e^{i\vec{k}\cdot\vec{R}_i} \hat{a}_k \qquad (2)$$

$$b_j = \frac{1}{\sqrt{N}} \sum_{\vec{k}} e^{i\vec{k}\cdot\vec{R}_j} \hat{b}_k \qquad (3)$$

$$\begin{vmatrix} \varepsilon_{2pz} - E & -t(1+e^{-i\vec{k}\cdot\vec{a}_1}+e^{-i\vec{k}\cdot\vec{a}_2}) \\ -t(1+e^{i\vec{k}\cdot\vec{a}_1}+e^{i\vec{k}\cdot\vec{a}_2}) & \varepsilon_{2pz} - E \end{vmatrix} = 0 \qquad (4)$$

In the case of ideal structure model, graphene can select atomic orbital energy level $\varepsilon_{2pz} = 0$ as the reference point of π band energy $E(k)$, and the energy band of graphene atom can be obtained by unfolding the above formula.

$$\begin{aligned} E^2 &= t^2(1+e^{i\vec{k}\cdot\vec{a}_1}+e^{i\vec{k}\cdot\vec{a}_2})(1+e^{-i\vec{k}\cdot\vec{a}_1}+e^{-i\vec{k}\cdot\vec{a}_2}) \\ &= t^2\{3+2\cos(\vec{k}\cdot\vec{a}_1)+2\cos(\vec{k}\cdot\vec{a}_2)+2\cos[\vec{k}\cdot(\vec{a}_1-\vec{a}_2)]\} \end{aligned} \qquad (5)$$

In Formula (5),

$$\vec{k}\cdot\vec{a}_1 = (k_x, k_y)(\frac{3a}{2}, \frac{\sqrt{3}a}{2}) = \frac{3a}{2}k_x + \frac{\sqrt{3}a}{2}k_y \qquad (6)$$

$$\vec{k}\cdot\vec{a}_2 = \frac{3a}{2}k_x - \frac{\sqrt{3}a}{2}k_y \qquad (7)$$

$$\vec{k}\cdot(\vec{a}_1-\vec{a}_2) = \sqrt{3}ak_y \qquad (8)$$

After elementary trigonometric function and difference operations, graphene band can be got as shown in formula (9).

$$E_{\pm}(k_x, k_y) = \pm t\sqrt{3+2\cos(\sqrt{3}k_y a)+4\cos(\frac{\sqrt{3}k_y a}{2})\cos(\frac{3k_x a}{2})} \qquad (9)$$

The energy band of graphene can be obtained from the calculation formula, and the structure and characteristics of graphene nanoribbons can be analyzed with the help of computer software.

4. Structure and characteristic analysis of graphene nanoribbons

In order to verify the feasibility of graphene band calculation, aided by computer simulation method, the structure and characteristics of graphene nanoribbons has been analyzed, the analytical model can be referred to the ideal Schottky barrier graphene nanoribbons field effect transistor (SBFET) and MOS graphene nanoribbons (MOSFET) with Youngki Yoon et al,. The parameters of the characteristic analysis are ID and VG characteristics.

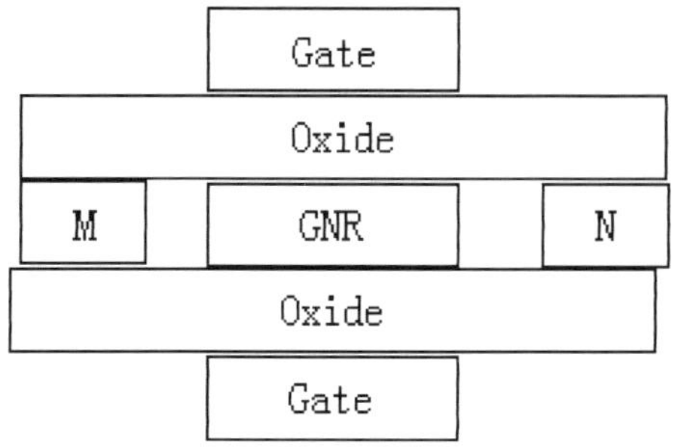

Figure 2. Schematic diagram of SBFET section

The section of the SBFET is shown in Figure 2, the channel material is graphene nanoribbons, and the source drain is contacted by Schottky. The results of MOSFET and SBFET profiles are similar. The difference is that the source and drain are doped. In the computer aided calculation mode, the SiO2 gate insulating layer thickness is 1.8nm, the dielectric constant is set to 3.9; the channel material n is set to 12, the length is 18nm, width is 1.38nm, the gap of Eg is about 0.8eV, the calculation results are shown in Figure 2 is obtained by calculating.

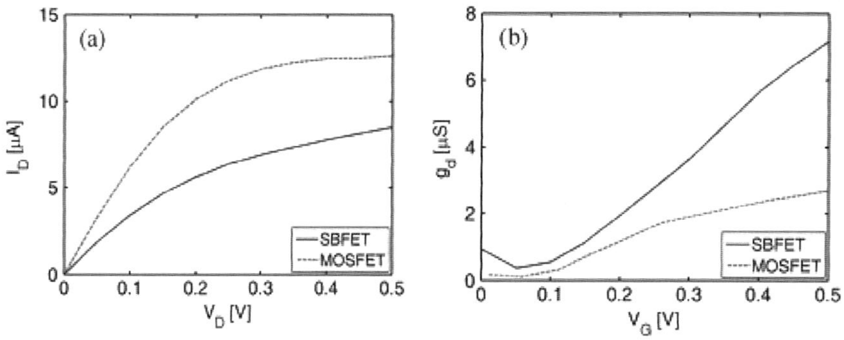

Figure 3. ID and VG characteristic curves

Figure 3 shows the VG and ID curve characteristics results obtained by computer aided calculation, the results can be seen from the saturated characteristics of MOSFET showed better characteristics, output from the electrical conductivity, smaller conductance MOSFET, it verified the feasibility of the calculation and analysis of field effect tube of graphene nanoribbons.

5. Conclusion

In order to open the band gap of graphene and realize its switching function, it is necessary to control the electrical properties of graphene by doping, which needs to analyze the structural characteristics of graphene. Taking graphene nanoribbons as an example, the ID and VG characteristics of Schottky barrier graphene nanoribbons field effect transistors (SBFET) and MOS graphene nanoribbons (MOSFET) were investigated by computer aided calculation. The analysis results show that MOSFET has more obvious advantages in terms of saturation characteristics and output conductivity. The research of graphene material is not mature, its experimental research is more complex, and the method of calculation and analysis can effectively shorten the research cycle and improve the efficiency of research. Therefore, it is of great practical significance to develop the calculation analysis method with the aid of the auxiliary calculation tools

6. Acknowledgements

The work was supported by the Youth Foundation Project of Wuhan Donghu University with the project number 2017dhzk002 and the project name Principle design of graphene superlattice superlattice field effect device.

7. References

[1] Chao Xu and Dong Wang 2014 An improved rough surface method to elastoplastic contact analysis model *J. Journal of Xi'an Jiao Tong University* **11** 25-26

[2] Cheng Zhang, Zhengqian, Luo Jin, and Zhang Wang 2012 Graphene double wavelength mode-locked Yb doped fiber laser with evanescent field of fused biconical optical fiber *J. Chinese Journal of Lasers* **6** 23

[3] Guodong Huang, Zifeng Ni and Da Bian. Effect of nano alumina on properties of metal matrix ceramic coatings *J. Journal of Northwestern University (Natural Science Edition)* **3** 9-11

[4] HuanTong, Guobao Jiang and Chen Yu 2014 Mechanically exfoliated graphene passive harmonic mode-locked erbium-doped fiber laser *J. Chinese Journal of Lasers* **8** 76-77

[5] Jun Ren, Sida Wu and Zhaochen Cheng 2015 Based on graphene oxide and semiconductor saturable absorption mirror mode-locked femtosecond erbium-doped fiber laser *J. Chinese Journal of Lasers* **6** 43-49

[6] Jiansheng Wang, Haixia Lu and Zhenwei Yan 2015 Steel surface electric spark deposition layer friction and wear performance *J. Transactions of Materials and Heat Treatment* **1** 33-35

[7] Li Zhang, Juntao Wu and Lei Jiang 2014 Graphene and its polymer nanocomposites *J. Progress in Chemistry* **4** 54-58

[8] Shaojuan Li, Sheng Gan and Haoran Mu 2014 Application of graphene optoelectronic devices in *J. New Carbon Materials* **5** 27-31

[9] Yuntao Xi, Wei Liu and Rui Liu 2015 Effect of preparation process on corrosion resistance of bimetal composite coating *J. Natural Gas and Oil* **6** 5-8

[10] Tao Ying, Zhuang Ma and Fanghong Sun 2014 Microstructure and Abrasive Resistance of glass-Ceramic Composite Coatings from Fly Ash and Al-TiO2-B2O3 *J. Materials for Mechanical Engineering* **9** 29-38

[11] Weihong Bi, Caili Li and Xiaoyu Wang 2016 Birefringence and electro optic control characteristics of graphene coated micro nano fibers *J. Acta Optica Sinica* **10** 3-5

[12] Weihong Bi, Yuanyuan Wang and Guangwei Fu 2016 Study on electro optic modulation characteristics of hollow fiber coated with graphene *J. Acta Physica Sinica* **4** 61-63

[13] Xuexing Ding, Ruqi Yan and Yonglei Jia 2014 Construction and analysis of fractal contact model of rough surface based on substrate length *J. Tribology* **4** 31-39

[14] Yongguang Wang, Zifeng Ni and Guozhong Chen 2013 Effect of Zinc Oxide adhesive in the ceramic coating content on Microstructure of coating *J. Materials Protection* **5** 60-61

[15] Zhuang Ma, Peng Zhou and Shizhi Dong 2013 Effect of preparation process of ceramic aluminizing composite coating on corrosion resistance of pure copper surface *J. Corrosion Science and Protection Technology* **4** 46-49

Preparation and Gas Sensing Properties of SnO_2 Hollow Nanostructures

Junhai Wang, Xiutao Ge and Yuping Chen

Material Science Chemical Engineering School, Chuzhou University, Chuzhou 239000, China

Abstract. Gas sensor can be used to detect, monitor, monitor, analyze and alarm. It has very important application value in industry, national defense, food safety and medical examination. Among the gas sensors, the semiconductor resistance type gas sensor has many advantages, such as high sensitivity, fast response, small size, light weight, easy to carry, and so on. As one of the widely used gas sensing materials, SnO_2 has been a hot spot in the research and application in 60s, and it has been developing rapidly. SnO_2 is one of the most common gas sensitive materials, with physical and chemical properties of gas detection and stability; reversible adsorption and desorption time is short; low cost, energy saving and other advantages, is widely used in the detection of various gas sensitive devices. However, the disadvantages of high working temperature and poor selectivity of SnO_2 make it need to be further improved as gas sensitive material. The gas sensing mechanism of semiconductor gas sensitive material is mostly controlled by the surface, which is embodied in: increasing the specific surface area of the material and enhancing the gas sensing performance. At present, the main method of increasing the specific surface area of the material is nano scale. In this paper, the preparation and gas sensing properties of porous and hollow structure SnO_2 are mainly discussed on the basis of particle size. Porous structure facilitates the increase of the specific surface area of the material, providing a larger surface area and volume ratio for the material, which is of great benefit to the diffusion and transportation of the gas being measured. Nano hollow structure has a thin shell; the inner and outer surface provides a high surface area and volume ratio, so that the sensor has a faster response and recovery speed, and higher sensitivity. Using carbon micro spheres as hard templates, the structure of SnO_2 was successfully prepared by using different raw materials ratio and hydrothermal conditions. The composition and morphology of SnO_2 were characterized by XRD, SEM, TEM and so on. The study of gas sensitivity shows that the SnO_2 of the hollow classification structure can show good gas sensitivity, which has the advantages of high sensitivity, good selectivity and strong stability.

1. Introduction

With the rapid development of modern industry, such as chemical industry, automobile, medical treatment, petroleum and so on, human beings are in possession of great wealth, but also bear the serious environmental pollution. A large number of harmful toxic gases (such as carbon monoxide, nitrogen oxides, sulfur oxides, methane and so on) in human production and life, all the time, are not a threat to our health. Not only that, the pollution of the climate caused by the gas, for the safety of industrial and agricultural production has also planted hidden dangers. And with the improvement of living standards, the pursuit of the increase of material life makes the car as a convenient means of transport has developed rapidly. But in the car to bring people living facilities at the same time, because of drunk driving and traffic accidents caused by a serious threat to people's lives and property safety. The appearance of the above phenomenon, so that the gas detection equipment came into being.

Nowadays, the detection and analysis of gas is more time-consuming and expensive testing instruments, such as spectral analysis, chromatographic analysis, and so on. Gas sensor has many advantages, such as good selectivity, high sensitivity, fast response time, short recovery time, good stability and low cost, which is widely used. The use of gas sensor can accurately monitor the toxic and harmful gases, effectively prevent the occurrence of accidents, and protect the air environment [1]. The gas sensor is closely linked with our daily life, such as ethanol concentration detection, detection of pesticide residue, gas storage, flammable gas detection based on gas sensor are realized. With the research, development and application of the gas sensor, it has been widely concerned by people and put forward a great challenge to the scientific research personnel. Gas sensor is an important branch of the sensor, the mechanism is to use a variety of physical effects, chemical reactions, and the measured gas needs to be detected in the extraction of parameters. The gas sensing principle of typical representative metal oxide semiconductor is based on the change of the conductivity of semiconductor gas sensing material with the change of ambient atmosphere. So far, a lot of semiconductor metal oxides, such as iron oxide, Zinc Oxide, copper oxide and cerium oxide and so have been used as gas sensing materials. SnO_2 as the most sensitive material widely, with physical and chemical properties of stability, gas detection, reversible adsorption and desorption time is short, low cost, energy saving and other advantages, has long been favored by researchers. SnO_2 is a kind of important wide band gap n type semiconductor material, which contains oxygen vacancy or tin gap ion. The gas sensing effect is obvious. It is generally believed that the gas sensing mechanism is the control mechanism of surface adsorption. Because SnO_2 has a special crystal structure that is structural defects, high surface activity, large surface area and adsorption characteristics, so that their gas sensing performance compared with the traditional sensor showed a unique performance[2].

2. Experimental test methods

Scanning electron microscopy was used to observe and analyze the microstructure of the surface of the material. The block or powder particles are bonded to the conductive adhesive, and no conductive samples need to be sprayed with carbon or gold. The prepared samples were put on the test bench for observation. Observation of high magnification and high resolution by scanning electron microscope. Scanning electron microscopy is formed by electron beam through electronic grid scanning at the surface of the sample, the aggregation of electron beam and sample produced by the interaction of the two electron emission quantity with the specimen surface changes, two electronic signals are converted into electrical signals after being input to the kinescope grid, finally reflect the specimen surface the two electron image. Differential thermal analysis (DTA) is a thermal analysis method to determine the physical and chemical properties of the material by measuring the temperature difference between the sample and the reference. The sample in the heating occurs in the process of solidification, melting, decomposition, synthesis, crystal transformation, desorption, adsorption and other physical or chemical changes caused by variations in enthalpy, thermal effect, the temperature difference between the environment and the system. The heat weight method (TG) is based on the material in the process of heating, chemical changes will occur, which will be accompanied by changes in temperature. The change process can be studied by measuring the change of material quality. Transmission electron microscopy observation is often used to hollow structure material, its working principle: when the electron beam and the thin specimen interaction will inspire and sample can be used for imaging of electron beam scattering and transmission electron beam. The transmission electron beam through a lens focused like called bright field scattering, electron beam through lens focusing the image called dark field image. After the multi stage amplification, the image to be observed is presented on the screen. The principle of semiconductor gas sensor is based on the chemical reaction between adsorbed gas molecules and sensitive material. For the N type semiconductor, such as the SnO_2, the resistivity change mechanism and process: when the different morphologies of SnO_2 gas sensitive material preparation components when exposed to air, the molecular surface redox reactions occur with air, the oxygen molecules in the air conduction band won the gas sensitive materials in the form of electronic chemical adsorption oxygen (such as O2-, O-and and O2-), which can reduce the carrier -SnO2 the number of electrons, the electron mobility is reduced, caused by decreased SnO2 conductivity. This system uses the voltage and current test method, the basic test principle is shown in figure 1. The characteristic of the gas sensor is reflected by the voltage on the load resistance R1 which is connected in series with the gas sensing element. The circuit will be a known resistance load resistance R1 and gas sensor series access total voltage is 5V, is to change the relationship between voltages of gas sensor with time by measuring the load voltage Vout resistance under different conditions. When the heating wire is provided with the working temperature of the gas sensor, the different working temperature can be obtained by adjusting the different heating voltage[3].

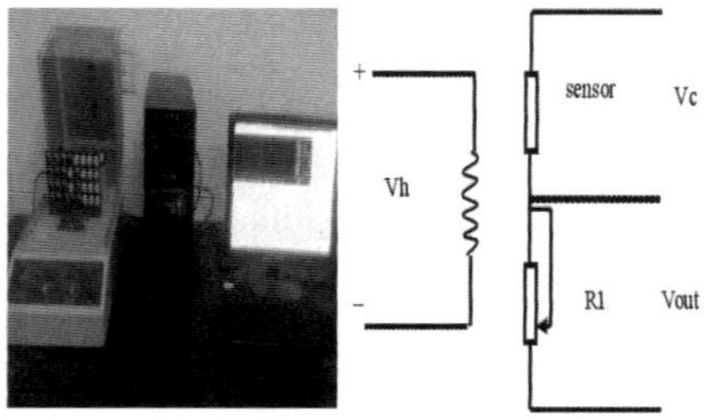

Figure 1. Schematic diagram of experimental device

3. Copyright form

3.1. Preparation of porous structure by FA- template method SnO_2

The high porosity and high specific surface area of porous micro / nano materials show the superior physical and chemical properties compared with the bulk structure of the corresponding materials. These properties play an important role in photo catalysis, gas sensitivity, solar cells, lithium ion batteries and drug transport. Therefore, the development of the preparation of porous structure is a hot research topic in recent years. The preparation of porous materials by direct introduction of structural additives and the preparation of porous materials have been successfully achieved. Such as latex spheres, oxalic acid erosion, emulsion, block copolymers, surfactants, and so on, have played an important role in the process of preparing porous structures through crystallization and precipitation. Furfuryl alcohol (FA) as a chemical material, wide source. Research shows that in the acid (as Lewis acid) under the action of furfuryl alcohol can be catalyzing the formation of furfural resin (FAR). Heating in the air, the FAR will have a continuous reaction of polymerization carbonization combustion. Many metal cations can play a role in the Lewis acid, according to this principle, can the design and synthesis of the three-dimensional porous structure of metal oxide by furfural method. In this study, we used $SnCl_2.2H_2O$ as the Lewis acid to catalyze FA to get a mixture of SnO_2-FAR, and then the porous structure was synthesized by combustion reaction control SnO_2. This preparation scheme, both as fuel and furfuryl alcohol conversion template. The preparation of porous structure has the advantages of furfuryl alcohol is easy to get raw materials, low cost and simple method, can be synthesized with high specific surface area and porosity of the porous material. Thermal analysis of porous structure SnO_2 by FA- template method is illustrated in Figure 2.

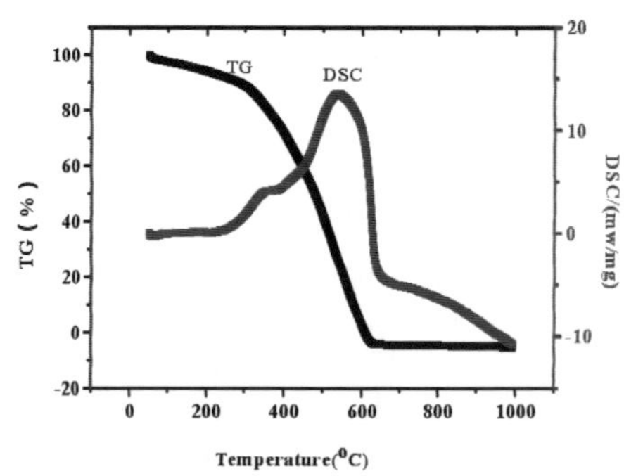

Figure 2. Thermal analysis of porous structure SnO_2 by FA- template method

3.2. Preparation of porous structure SnO_2 nano cubic block by phase separation method

Phase separation refers to the two kinds of incompatible oxide phase (the main frame material and the hole forming body) after mixing calcined into dense matter, then one of the constituents of selective eroded by chemical method or evaporation, thus leaving the equivalent gap in the main frame material. The porous metal oxides prepared by this method are MnO, TiO_2, TiN, ZrO_2 and so on. The phase separation method is proved to be an effective method to control the porous structure. In this paper, we use a simple wet chemical method, and prepared by poly - precipitation nano cubic precursor of CoSn $(OH)_6$ (CHS), and then calcining the mixture of Co_3O_4 and SnO_2, and finally get the pure porous structure of SnO_2 in pickling conditions. The whole reaction is the process of decomposition, aggregation and dissolution. For the selection of the best operating temperature of the test, we selected the measurement temperature range of 260-400. The sensitivity curve is drawn as shown in Figure 3. From the graph we can see that, with the increase of temperature, the sensitivity of the porous structure SnO_2 gas sensitive material to 200 ppm ethanol is gradually increased, and the sensitivity reaches the maximum value of 105 when the temperature reaches 300. In 300 C, with the increase of temperature sensitivity decreased linearly, so we selected 300 DEG C for optimum temperature response of gas sensors to ethanol gas, the temperature will also serve as a back test of different concentrations, different gas temperature[4].

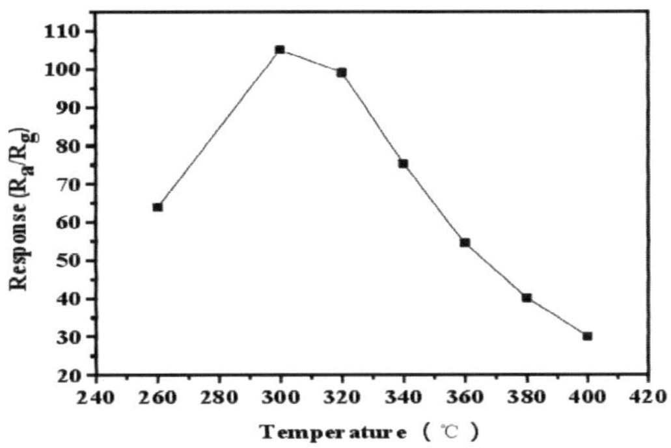

Figure 3. Sensitivity curve of different temperature

3.3. Advantages of self-template method

Since the template method is by chemical erosion, Ostwald aging, Kirkendall effect mechanism will continue to consume the precursor template internal reaction, direct the formation of hollow structure. In this process, the precursor template is both a reactant and a product, without additional removal of the template. Hollow structure has been developed so far, the hollow particles prepared by the simple method are spherical and polycrystalline structure, because the non-spherical template is difficult to obtain, so people are still working hard to prepare non spherical single crystal particles. Wang et al have been using the self-template method to get the polyhedron Cu_2O hollow lattice by using the single crystal Cu_2O template. Preparation of non-spherical SnO_2 is very difficult, it is necessary to synthesize the non-spherical precursor under the aid of other metal elements, and then the introduction of the metal phase chemical erosion, the pure phase of non-spherical porous SnO_2. Due to the particle surface with free surface of high energy, so the reaction activity and reaction rate at the top corner of the large, so large vertex curvature is first dissolved, whole. Once a hole is formed, the OH- passes through the hole to penetrate into the inside of the cube. As the cube ZHS external than internal stability, so continue to carry out the decomposition reaction inside. Eventually, with the continuous generation and release of Zn (OH) 42- and 62- (Sn) OH, the formation of cubic hollow ZHS. The formation of the hollow structure can be understood: the central part of the interior cube, is in the raw material concentration and unbalanced conditions through the rapid precipitation polymerization reaction to form, as a result of the growth is relatively fast, so the structure is not stable, is in a metastable state.

4. Experimental result analysis

Through the electron microscope, it can be observed that there are a large number of solid fourteen surface ZHS generation, fourteen ZHS particles of uniform size, complete structure, smooth surface and no defects. The dispersion of particles is not very good, is often a few particles together to grow. The cause of this phenomenon may be caused by uneven stirring during the preparation process. Fourteen surface body structure size in 0.8-1 m or so. Can generate fourteen surface morphology because, with the existence of excess OH-, increase of OH- adsorption on a surface, inhibiting crystal growth on the surface of the shaft in the vertical direction, so that the surface area increases, the surface morphology is formed of fourteen polyhedron. After the erosion of the ZHS precursor and the erosion of the same, still maintained a fourteen surface appearance, and the internal ZHS has been reacted with an excess of OH-, the formation of a hollow structure. Scanning electron microscope photographs of the broken fourteen surface can clearly see the shell structure of the hollow spherical shell, the thickness of the shell layer is about 10 nm, the thickness is uniform, the surface is smooth, and the hollow diameter is around nm 400-600. Taking ZHS as precursor, the cubic and polyhedron morphology of solid and hollow structure were successfully prepared by self-template method. The results of XRD test showed that the solid and hollow structures were pure phase ZHS. The results of scanning electron microscope can be analyzed, and the solid and hollow cubic shape diameter is about nm 150-200, and the wall thickness of the hollow cube is about 40 nm. The diameter of solid and hollow fourteen surface is about 0.8-1 m, and the wall thickness of the hollow fourteen surface is about 10 nm, and the diameter is about nm 400-600. NaOH plays an important role in the process from cube to polyhedron and from solid to hollow. However, after the calcination and pickling, the morphology is badly damaged, and it is very different from the expected results[5].

5. Conclusion

In this paper, the SnO_2 hollow microspheres with hierarchical structure were prepared by hydrothermal method, and the SnO_2 hollow structure of Au was obtained by a simple thermal decomposition process. To study the effects of hydrothermal time on the size and morphology of SnO_2 microspheres, the formation mechanism of the hierarchical structure of SnO_2 hollow microspheres may product are discussed; in the preparation of hollow microspheres and solid SnO_2 microspheres for comparison, study its sensitivity to ethanol gas; from the hollow cubic ZHS particles prepared by template the well dispersed, the particle diameter of about nm in 150-200, wall thickness of about 40 nm. Fourteen surface morphology of hollow 0.8-1 in diameter m, wall thickness of about 10 nm in diameter, hollow nm left 400-600 But did not generate the expected porous SnO_2. The yield of the hollow cubic CHS prepared by the self-template method is large, and the diameter of the particles is about 100 nm. But did not generate the expected porous SnO_2.

6. Acknowledgements

This paper is a natural science research project of Chuzhou University (2010kj003B) and a young talent foundation of Chuzhou University (2013RC014).

7. References

[1] Xu L, Chen W, Jin L and Zeng W 2016 A novel SnO_2 nanostructures and their gas-sensing properties for CO *J. Journal of Materials Science Materials in Electronics* **27** 4826-4832
[2] Liu J, Meng F, Zhong Y, Liu J, Chen G, Wan Y, Qian K and Sitaramanjaneya M.T 2000 Assembly, formation mechanism, and enhanced gas-sensing properties of porous and hierarchical SnO_2hollow nanostructures *J. Journal of Materials Research* **25** 1992-2000
[3] Noipa K and Pukird S 2010 Studies on the Preparation and Gas Sensing Properties of SnO_2 Nanostructures at Room Temperature *J. Advanced Materials Research* **93-94** 227-230
[4] Bhardwaj N, Pandey A, Satpati B, Tomar M, Gupta V and Mohapatra S 2016 Enhanced CO gas sensing properties of Cu doped SnO_2 nanostructures prepared by a facile wet chemical method *J. Physical Chemistry Chemical Physics Pccp* **18** 18846
[5] Mustapha W.N.W, Abdul Hamid S. A. R. S, Hutagalung S. D, Hieu N. V, Mohamed K and You C. K 2013 Synthesis and Gas Sensing Properties of SnO_2 Nanostructures by Thermal Evaporation *J. Advanced Materials Research* **62** 350-355

IOP Conf. Series: Materials Science and Engineering **422** (2018) 012010 doi:10.1088/1757-899X/422/1/012010

Germanium Atom Substitution in Monolayer Graphene:A First-principles Study

Muhammad Rafique[(a),b)*)], Shuai Yong[(b)], Irfan Ali[(c)] and Irfan Ahmed[(a)]

[(a),*)]Mehran University of Engineering and Technology, S.Z.A.B, Campus Khairpur Mir's, Sindh, Pakistan
[(b)]School of Energy Science and Engineering, Harbin Institute of Technology,92 West Dazhi Street, Harbin 150001, PR China
[(c)]Institute of Chemistry, Shah Abdul Latif University, Khairpur Mirs', Pakistan

E-mail:shuaiyong@hit.edu.cn

Abstract. In this work, structure, electronic and optical parameters of germanium (Ge) atom substituted monatomic graphene are demonstrated through first-principles study (FPS) computations. The concentration of Ge atoms was changed from 2.5 % to 7.5 % and the effects of varying concentration on aforementioned properties were investigated. It is observed that, replacing C atoms with Ge in graphene leads to a finite bandgap opening at the Dirac K-point, thereby producing a direct bandgap semiconducting graphene. We also found that, Ge doping in graphene significantly changes its refractive index parameter. Moreover, Ge atom doping in graphene reduces the overall absorption coefficient, though it observes a considerable red-shift towards the visible region of spectrum. Graphene reflectivity improves in low lying energy region after Ge atom substitution in its lattice. These results can pave a new route for tuning the electronic and optical properties of graphene to make it functional for nanoelectronics and optoelectronic device applications.

1. Introduction

Ever since the first report on graphene in 2004 [1], a sufficient amount of work and time has been attributed by the researchers to manipulate the intrinsic properties of this fascinating material to make it useful for real engineering applications. The electronic structure of graphene can be modified by chemically doping or adsorbing foreign atoms onto graphene. These foreign atoms can work as accepter or donor impurities in the graphene structure which can move the Fermi energy level (E_F) up or down depending upon the nature of impurity atoms/clusters [2,3]. Some studies [4,5] reveal that semiconducting graphene can be obtained, when nitrogen (N) or boron (B) atoms are incorporated in its lattice. Since graphene contains zero bandgap but Ge atom is a wide bandgap semiconductor; Hence substitution of Ge atom in graphene is a natural proposition for producing an ample and well-defined bandgap in monolayer graphene [6,7].

It is well known that, pure graphene absorbs 2% of light in the visible region [8]. Functionalizing graphene for absorbing specific wavelength of spectra, it is crucial to alter the absorption spectrum of graphene to make it practical for optoelectronic devices. Sedelnikova et al. [9] and Marinopoulos et al. [10] have comprehensively examined the optical properties for rippled graphene and flat boron nitride (BN) sheet, respectively. Some FPS calculations have been performed on the electronic and optical characteristics of intrinsic and extrinsic graphene and graphynes [11-16]. These studies reveal that, doping hetro atoms in graphene and graphynes can transform the intrinsic parameters of host materials.

Very recently, M. L. Ould Ne et al. [11] performed some FPS calculations on Ge atom-doped graphene structures. Authors reveal that, by introducing Ge atom in graphene lattice a well-defined bandgap in graphene electronic structure can be achieved. Though, in this study the concentration of host C atoms is varied while dopant impurity is kept constant. Hence, in this paper we try to explore the effects of variation in the concentration of impurity atoms on the intrinsic properties of graphene while keeping the supercell size as constant.

2. Details of computation

DFT calculations are performed on Ge atom-doped graphene structures, through GGA approach implemented in Vienna Ab-initio simulation package (VASP) [17,18]. VASP utilizes PAW potentials [19] with PBE functional [20] of exchange-correlation. A 450 eV kinetic energy cut-off was utilized for the expansion of wave function. A Graphene 5 × 4 supercell structure was used in this work with varying concentration of Ge atoms present in its lattice. 15 Å vacuum layer thickness in the Z direction is applied in order to eradicate the interaction between nearby layers. For Brillouin zone (BZ) sampling, we adopted Γ-centered 9 × 9 × 1 k-point grid. Given structural models were relaxed completely until less than 0.01 eV/Å Hellmann–Feynman forces and the energy change less than 10^{-6} eV was gained, respectively. To satisfy partial occupancy problems we utilized Gaussian smearing method.

3. Results and discussions

3.1. Formation energies of Ge doped graphene

Reliability of all Ge atom-doped graphene systems can be verified by calculating the formation energy; which can be obtained through following expression [21],

$$\mathbf{E}_{(\mathbf{gpGe})} = \mathbf{E}_{(\mathbf{gpGe})} - \mathbf{E}_{\mathbf{pgp}} + \mathbf{m}\mu_{\mathbf{c}} - \mu_{\mathbf{Ge}} \quad \mathbf{(1)}$$

here, terms $\mathbf{E}_{\mathbf{pgp}}$ and $\mathbf{E}_{(\mathbf{gpGe})}$ are the total energies of pure and Ge atom-doped graphene structures, respectively. The chemical potentials of carbon $\mu_{\mathbf{c}}$ and germanium $\mu_{\mathbf{Ge}}$ were gained through pristine graphene and standard phase of Ge crystal structure [22]. The **m** represents number of C atoms replaced by **n** number of Ge atoms in its lattice. The formation energy values of -0.96 eV, -2.83 eV and -3.96 eV were obtained for 1 Ge, 2 Ge and 3 Ge substituted graphene complexes, respectively. The negative formation energies suggest that, Ge atom substitution in graphene lattice is thermodynamically favorable. The trends of the obtained formation energies of Ge doped graphene systems are consistent with earlier available reports [23].

3.2. Structural and electronic parameters of Ge substituted graphene

In this work, we investigated the effects of Ge atom substitution on aforementioned parameters. Ge atoms with varying concentration of 2.5% (1 Ge atom), 5% (2 Ge atoms) and 7.5% (3 Ge atoms) were incorporated in graphene lattice containing 40 host C atoms, respectively. Optimized geometry of 1 Ge, 2 Ge and 3 Ge atom-doped monolayer graphene is presented in Figs. 1(a)-(c), respectively. Due to larger difference between covalent radius of Ge (120 pm) and C atom (73 pm), the increase in the lattice constant of graphene occurs. Larger covalent radius of Ge atom causes local deformation in the graphene lattice, which in turn produces variation in the C-C atom bond length. In order to investigate the variations in C-C atoms bond length after Ge atom substitution, the C atoms present around the Ge impurities were not fixed, rather, were allowed to settle down during geometry relaxation.

Given Fig. 1 presents the bond distances for Ge-C and C-C constituent atoms, it is found that, during geometry relaxation process the 2D planar structure of the graphene was retained after Ge atom substitution in its lattice. Ge atoms form strong covalent bonds with adjacent C atoms via sp^2 hybridization process. Constituent C-C atom bond distances present in vicinity of Ge atoms were calculated in variety of 1.27-1.36 Å and obtained disparity in C-C atoms bond length was in variety of 0.14 Å to 0.06 Å. Bond distances of Ge-C atoms were obtained in range of 1.61-1.67 Å. In addition, Ge atom substitution produced Jahn-Teller distortion in graphene lattice by lowering the D_{3h} local symmetry of graphene to C_{3v}, C_s or C_i symmetry during relaxation process. This calculation is in consensus with statement that, three equatorial bond distances of Ge-C atoms are different [16, 24].

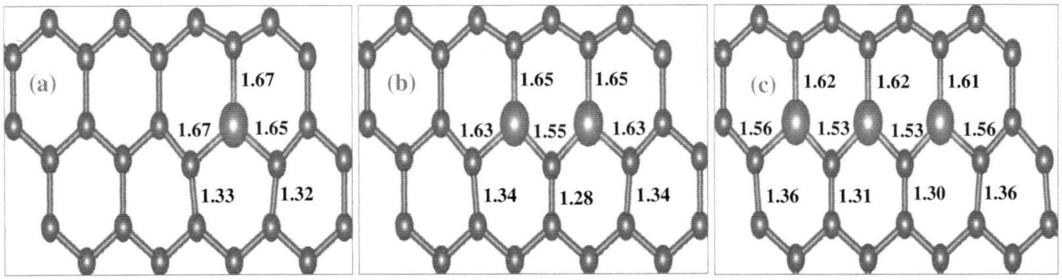

Figure 1. Top views of atomic structures Ge doped graphene systems showing bond length of Ge–C and C–C atoms in Å.

Next, the electronic structure of individual varying concentration of Ge atom-doped graphene systems is investigated. Band structure diagrams were obtained for all Ge doped graphene complexes. 30 *k*-point were adopted for each symmetrical line through path Γ- M - K - Γ in irreducible BZ for electronic structure calculation with refined grid.

Different Ge (1 Ge, 2 Ge and 3 Ge) atom-doped graphene system band structure diagrams are described in Figs. 2(a)-(c), respectively.

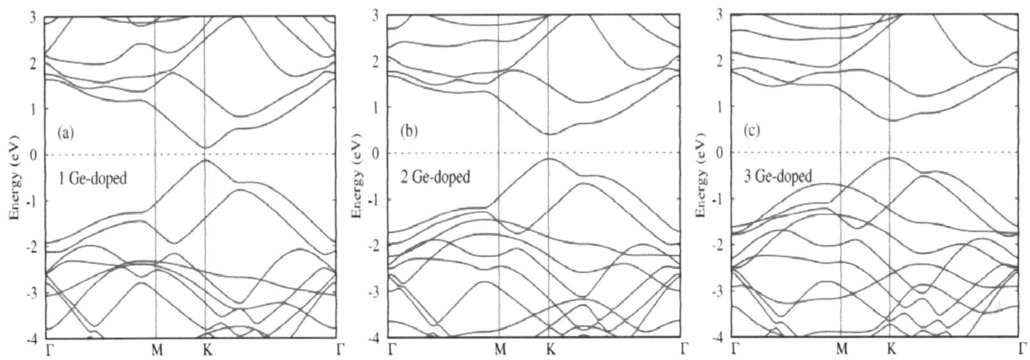

Figure 2. Electronic structure of Ge atom-doped graphene (5 × 4) supercell systems.

As we know that, Ge impurity atoms carry extra charge carriers than host C atom, so the valence electrons of Ge atoms hybridize with π^* electrons of graphene. Since Ge atoms form covalent bonding with C atoms of graphene, which causes the disruption in the symmetric structure of graphene, thereby inducing a finite band gap in its electronic structure [7, 25]. Moreover, the bandgap value induced in the electronic structure of graphene is in direct proportion to number of Ge atoms present in its lattice, which is also evident in the electronic structures shown in Figs. 2(a)-(c), respectively. Single Ge atom substitution in graphene introduced a bandgap of ~0.22 eV at the Dirac point, that is consistent with earlier reports [25, 26]. When the concentration of Ge atoms present is graphene lattice is increased to 5% (i.e. 2 Ge atoms) and7.5% (i.e. 3 Ge atoms), the bandgap value increases to ~0.8 eV and ~1.1 eV, respectively. Obtained results suggest that an ample and well-defined bandgap in graphene electronic structure can be achieved by doping Ge atoms in its lattice. During Ge atom substitution in graphene, the π band of graphene stands firm at approximately -0.03 eV energy, although π^* band creeps into conduction band and away from the (E_F) level depending upon the number of Ge atoms present in graphene lattice. This behaviour suggests that, the Ge atom p orbital electrons add their energy into conduction band, thereby altering the position of conduction band minimum (CBM). By increasing the concentration of dopant atoms, some extra energy is added to CBM, which in turn modifies the position of CBM in the electronic structure, thus increasing the existing energy difference between the valence band and conduction band [11]. These predictions are consistent with the previous reports [26-28].

3.3. Optical parameters of Ge substituted graphene

Finally, we investigated the optical parameters of Ge atom-doped graphene systems through Random Phase Approximation (RPA) [29] technique within DFT scheme. Local field effects are omitted and interband transitions are included in this method. Hence, some inaccuracies in dielectric constant can be expected at low energies. To calculate the refractive index, absorption coefficient and reflectivity plots, we require dielectric constant. Since, dielectric constant is summation of real and imaginary parts i.e. $\varepsilon = \varepsilon' + \mathbf{i}\varepsilon''$ together. From dielectric constant, we can simply extract the refractive coefficient 'n', extinction coefficient 'k', absorption coefficient 'α' and reflectivity 'R'. Calculation technique for these optical properties is demonstrated in detail in Ref. [10]. Our obtained optical parameters for pure graphene are consistent with previous reports [8, 9, 30].

Figs. 3(a) and 3(b) present the refractive index plots for pure and Ge atoms substituted graphene structures and the comparison is also provided. Static refractive index (i.e. value of refractive index at zero energy) for pure graphene is found as 2.75, and its smallest peak appears at 5 eV energy. Though, after Ge atom substitution, static refractive index value is increased. During single Ge atom substitution, static refraction index (n) value increases to ~3.6 as given in Fig. 3(a).

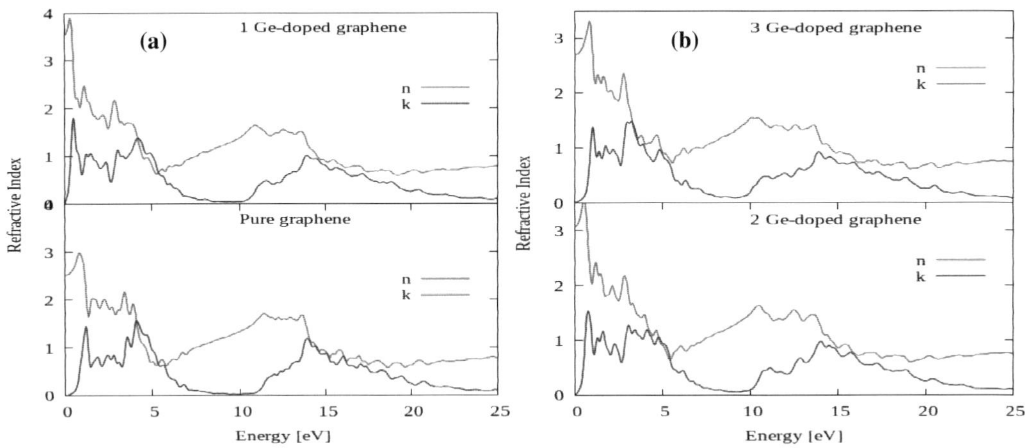

Figure 3. Refractive Index of (a) Pure, 1 Ge atom-doped and (b) 2 and 3 Ge-atom doped graphene

Likewise, pure graphene extinction coefficient (k) maximum peaks emerge at 1 eV and 5 eV energy with 1.4 and 1.6 peak intensities. However, after single Ge atom-doping, the first maximum peak shifts to lower energy at 0.7 eV with 1.85 peak intensity. This behaviour suggests that Ge atom substitution can produce shift in the extinction coefficient towards lower energy and also produces increment in the peak intensity of extinction coefficient as shown in Fig. 3(a), respectively. When concentration of Ge atoms is increased to 5% and 7.5%, similar changes in static 'n' and 'k' values as mentioned above are achieved as described clearly in Fig. 3(b). To sum up, one can assume that Ge atom incorporation in graphene can significantly change the refractive coefficient (n) and extinction coefficient (k) parameters of graphene.

Pure and Ge atom substituted graphene system absorption coefficient plots are presented in Fig. 4(a), respectively. Two major absorption peaks emerging at ~4.5 eV, ~14 eV energy levels having intensity of ~3500 cm^{-1} and 8700 cm^{-1} are observed in pure graphene absorption spectrum as evident in 4(a), respectively [31]. The first peak at 4.5 eV is associated with $\pi \rightarrow \pi^*$ transitions and the second peak appearing at 14 eV is associated with $\sigma \rightarrow \sigma^*$ transitions. Moreover, zero absorption coefficient quantity is obtained in the energy range of 0-0.5 eV and also from 7-11 eV for pure graphene system [8, 32]. However, when Ge is embedded in graphene, it enables absorption coefficient parameter to start from ~0 eV energy, indicating, Ge doping can produce red shift in the absorption spectrum of pure graphene which is evident in 4(a). Similarly, some changes in the absorption quantity values are also achieved in 7-11 eV energy intervals. During Ge atom incorporation in graphene, highest absorption peak attributed to σ→σ* transitions appearing at 14 eV for pure graphene attains lesser values and the smallest absorption peak attributed to π→π* appearing at ~4.5 eV is drifts towards low energy range. Through obtained absorption coefficient parameter, one can predict that Ge atom incorporation in graphene, reduces the absorption quantity as a whole; Though, provides a red-shift in absorption spectrum towards visible region.

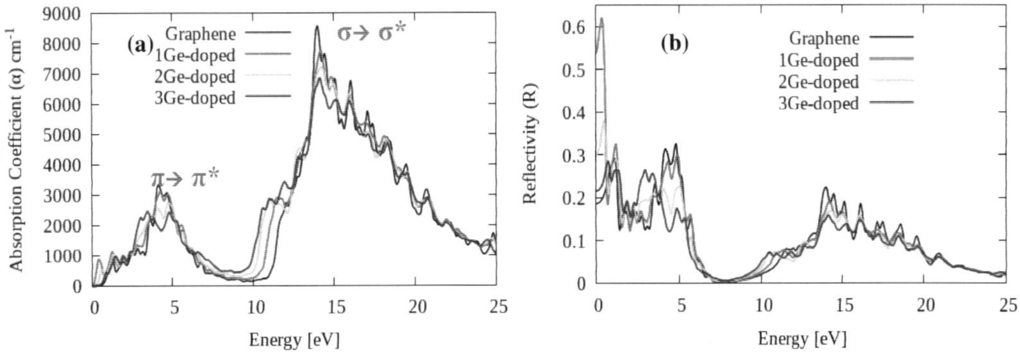

Figure 4. (a) Absorption coefficient and (b) Reflectivity plot of pure and Ge substituted graphene systems

Fig. 4(b) describes the reflectivity diagram for pure and Ge substituted graphene systems. The reflectivity plot of intrinsic graphene contains three main peaks with ~0.29, 0.32 and 0.21 intensities at ~0.7 eV, ~4.5 eV and ~14 eV energy levels as presented in Fig. 4(b), respectively [8, 32]. However, when Ge atoms are incorporated into graphene, the two main peaks emerging at ~4.5 eV and ~14 eV present reduced intensity values as the concentration of Ge atom is increased which is evident in Fig. 4(b). Additionally, the low lying energy reflectivity peak of pure graphene appearing at 0.7 eV with 0.29, intensity has increased intensity of approximately 0.62 and 0.38 for 1 and 2 Ge atom-doped monolayer graphene systems as given in Fig. 4(b), respectively. When Ge atom concentration reaches to 7.5 %, the overall reflectivity parameter shows diminishing behaviour. It can be concluded that, introduction of Ge atoms in graphene lattice, provides increase in the static reflectivity (R at 0 eV) while the reduction in reflectivity in higher energy region is achieved as presented in Fig. 4(b), respectively [8, 32].

4. Conclusions

The structural, electronic and optical parameters for Ge doped graphene structures were investigated using FPS calculation based on DFT method. The concentration of Ge atoms was varied from 2.5 % to 7.5 % and their effects were analyzed. It is revealed that, Ge atom incorporation in graphene makes it display direct bandgap semiconductor with bandgap value of ~0.22 eV, ~0.8 eV and ~1.1 eV for 1 Ge, 2 Ge and 3 Ge atom doped graphene systems, respectively. For optical parameter calculations, it is found that, n and k peak intensities have increased values after Ge atoms substitution in graphene lattice. It is also found that, the minimum absorption peak associated to π→π* transitions appearing at 4.5 eV energy in graphene absorption spectrum exhibits a red shift in between ~0.1-0.3 eV energy towards visible range of spectrum when Ge atoms are placed in its lattice. Also the absorption coefficient appears at 0 eV energy after Ge atom doping. Similarly, reduced reflectivity parameter in higher energy region is gained while higher peak intensities in the lower energy region are obtained after Ge atom substitution into graphene. In summary, we can suggest that, Ge incorporation in graphene can produce direct bandgap semiconducting graphene. Moreover, Ge atom doping in graphene produces new trends in its optical absorption. These results pave a new route for graphene to be functional in the fields of photonics and nanoelectronics.

5. Acknowledgments

This work was supported by the Higher Education Commission, Pakistan under SRGP (No:21-1778/SRGP/R&D/HEC/2017). In addition, we would like to acknowledge the support that NVIDIA provided us through the GPU Grant Program.

6. References

[1] K. S. Novoselov, A. K. Geim, S. V. Morozov, D. Jiang, Y. Zhang, S. V. Dubonos, I. V. Grigorieva and A. A. Firsov, science **306** (5696), 666-669 (2004).
[2] P. Rani and V. Jindal, RSC Advances **3** (3), 802-812 (2013).
[3] R. Nascimento, J. d. R. Martins, R. J. Batista and H. Chacham, The Journal of Physical Chemistry C **119** (9), 5055-5061 (2015).
[4] L. Ci, L. Song, C. Jin, D. Jariwala, D. Wu, Y. Li, A. Srivastava, Z. Wang, K. Storr and L. Balicas, Nature materials **9** (5), 430-435 (2010).
[5] T. Martins, R. Miwa, A. J. Da Silva and A. Fazzio, Physical review letters **98** (19), 196803 (2007).
[6] X. Miao, S. Tongay, M. K. Petterson, K. Berke, A. G. Rinzler, B. R. Appleton and A. F. Hebard, Nano letters **12** (6), 2745-2750 (2012).
[7] P. A. Denis, ChemPhysChem **15** (18), 3994-4000 (2014).
[8] P. Rani, G. S. Dubey and V. Jindal, Physica E: Low-dimensional Systems and Nanostructures **62**, 28-35 (2014).
[9] O. Sedelnikova, L. Bulusheva and A. Okotrub, The Journal of chemical physics **134** (24), 244707 (2011).
[10]. A. Marinopoulos, L. Reining, A. Rubio and V. Olevano, Physical Review B **69** (24), 245419 (2004).
[11] M. O. Ne, A. Abbassi, A. Benyoussef, H. Ez-Zahraouy and A. El Kenz, Optical and Quantum Electronics **49** (6), 218 (2017).
[12] S. Chopra, ChemPhysChem **16** (9), 1948-1953 (2015).
[13] S. Chopra, RSC Advances **6** (25), 20565-20570 (2016).
[14] S. Chopra and L. Maidich, RSC Advances **4** (92), 50606-50613 (2014).
[15] R. Muhammad, Y. Shuai and H.-P. Tan, Physica E: Low-dimensional Systems and Nanostructures (2016).
[16] J. Liu, Z. Xin, X. Yan, H. Li and M. Yu, Solid State Communications **258**, 38-44 (2017).
[17] G. Kresse and D. Joubert, Physical Review B **59** (3), 1758 (1999).
[18] G. Kresse and J. Furthmüller, Computational Materials Science **6** (1), 15-50 (1996).
[19] P. E. Blöchl, Physical Review B **50** (24), 17953 (1994).
[20] J. P. Perdew, K. Burke and M. Ernzerhof, Physical review letters **77** (18), 3865 (1996).
[21] D. Li, C. Wang, Y. Niu, H. Zhao and C. Liang, Chemical Physics Letters **601**, 16-20 (2014).
[22] M. Rafique, Y. Shuai, H.-P. Tan and H. Muhammad, Applied Surface Science **408**, 21-33 (2017).
[23] P. A. Denis, Computational and Theoretical Chemistry **1097**, 40-47 (2016).
[24] E. Beheshti, A. Nojeh and P. Servati, Carbon **49** (5), 1561-1567 (2011).
[25] M. S. Azadeh, A. Kokabi, M. Hosseini and M. Fardmanesh, Micro & Nano Letters **6** (8), 582-585 (2011).
[26] E. Aktürk, C. Ataca and S. Ciraci, Applied Physics Letters **96** (12), 123112 (2010).
[27] T. Cui, R. Lv, Z.-H. Huang, S. Chen, Z. Zhang, X. Gan, Y. Jia, X. Li, K. Wang and D. Wu, Journal of Materials Chemistry A **1** (18), 5736-5740 (2013).
[28] Z. Liu, J. Li, Z.-H. Sun, G. Tai, S.-P. Lau and F. Yan, ACS nano **6** (1), 810-818 (2011).
[29] M. Gajdoš, K. Hummer, G. Kresse, J. Furthmüller and F. Bechstedt, Physical Review B **73** (4), 045112 (2006).
[30] A. Laref, A. Ahmed, S. Bin-Omran and S. Luo, Carbon **81**, 179-192 (2015).
[31] M. Rafique, Y. Shuai, H.-P. Tan and M. Hassan, RSC Advances **7** (27), 16360-16370 (2017).
[32] R. Muhammad, Y. Shuai and H.-P. Tan, Physica E: Low-dimensional Systems and Nanostructures **88**, 115-124 (2017).

IOP Conf. Series: Materials Science and Engineering **422** (2018) 012011 doi:10.1088/1757-899X/422/1/012011

The Application of Phase Field Method in Microstructure Evolution Simulation

C Y Teng[1], M Li[1], Z Y Zeng[1] and Z Y Ren[1]

[1] China Aero-Polytechnology Establishment, 7 Jingshun Road, Beijing 100028, China

E-mail: teng_cy@126.com

Abstract. The physical properties and mechanical behaviour of metals are determined by their microstructure frequently. The microstructure is referred to the size, shape and spatial arrangement of phases, grains, etc. These microstructure features have the mesoscopic length scale. How to obtain the desired microstructure is one main task of metal alloy design. However, it is difficult because the microstructure evolution is influenced by the chemical driving force during phase transformation, the composition of the alloy, etc. With the development of computing power and computer modelling, the computational material science is playing an important role in the microstructure design. Especially the kinetic phase field method is a powerful tool which is fast developed and widely used in these years. The phase field method is a phenomenological approach based on classical thermodynamic and kinetic theories. It describes a microstructure by using field variables. In this paper, the basic principle and application of this method are presented.

1. Introduction

The metal alloy design has started from thousand years ago. However, the influence of microstructure on materials' properties is known until hundred years ago. Nowadays, the microstructure design of metal alloy is popular. Especially, the microstructure tailor through computational materials science is a vital method to obtain the required properties of materials.

The computational materials science is quickly developed in these years due to the fast improvement of the computing power. The phase field method which is one of the computational material sciences has shown strong power in the microstructure design from the beginning of its development.

The phase field method is a phenomenological approach based on classical thermodynamic and kinetic theories. It describes a complicated microstructure (chemical and structural non-uniformities) as a whole using a set of field variables that are spatially continuous and time dependent [1-5]. In the following sections, the basic principle and application of this method's will be presented.

2. Principle of the phase field method

The spatial distribution and time-evolution of the microstructure can be described by non-conserved structural order parameter fields, $\eta_p(r, t)$ as a function of position and time, where $p = 1 \sim n$. The number n is determined by the structural characters. For example, the system shown in figure 1, there are two phases, i.e., the red phase on the left and the blue phase on the right. Therefore, two variants are needed to describe the microstructure evolution, h1r and h2r. The profile of these two variants along the Z direction through the central axis is shown in figure 2(a). The concentration profile is shown in figure 2(b). The growth process of variant h1r is shown in figure 2(c). It indicates that the red phase grows by consuming the blue phase. In brief, the microstructural evolution of alloys can be expressed by these two kinds of variants, i.e., the structural order parameter and concentration parameter. For example, the red phase indicates the water and the blue phase indicates the ice in Fig. 1,while the variants evolution shown in figure 2(c) means the melt of ice.

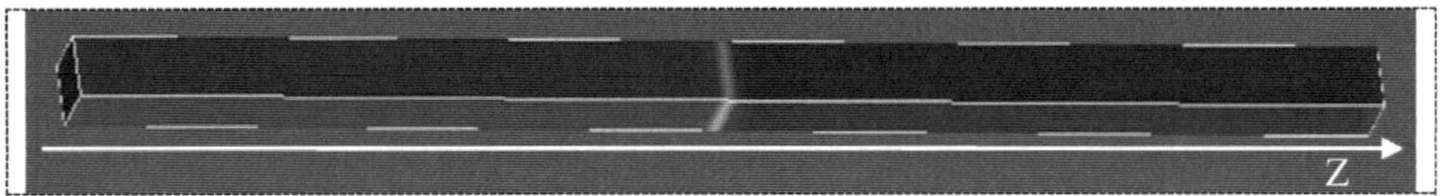

Figure 1. The red phase on the left and blue phase one the right in the system. The thin central layer is the interface of the two phases.

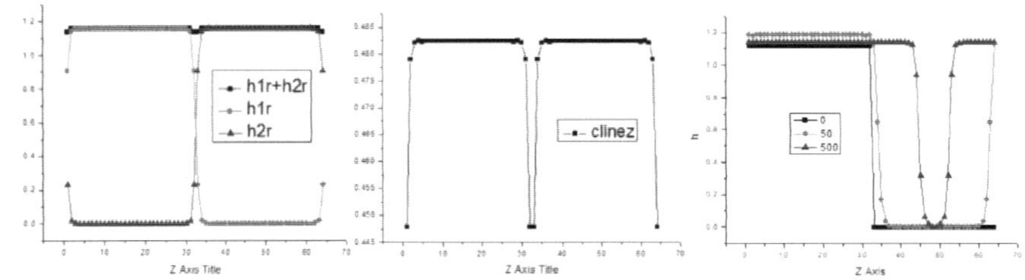

(a) (b) (c)

Figure 2. The structural order parameter profile (a) and concentration profile (b) of the two phases. The growth process of variant h1r (c).

The driving force of the microstructure evolution is the reduction of the total chemical free energy which consists of the local specific chemical free energy, the interfacial energy and the elastic strain energy. The local specific chemical free energy profile is shown in figure 3. It is a function of the structural order parameter and the concentration.

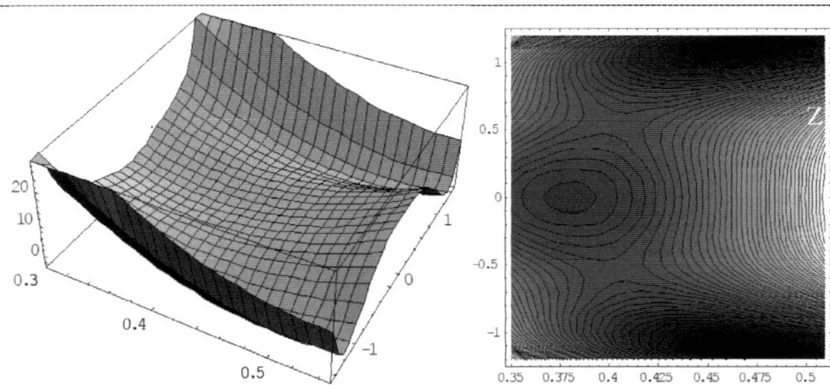

(a) (b)

Figure 3. The local specific chemical free energy profile, (a) 3D view plot and (b) top view plot. This energy is a function of the structural order parameter and the concentration

The chemical driving force of the phase transformation is the energy difference between the phase alpha and phase beta shown in figure 4. The alpha phase is stable when the concentration is less than about 0.4, while the beta phase is stable on the other side.

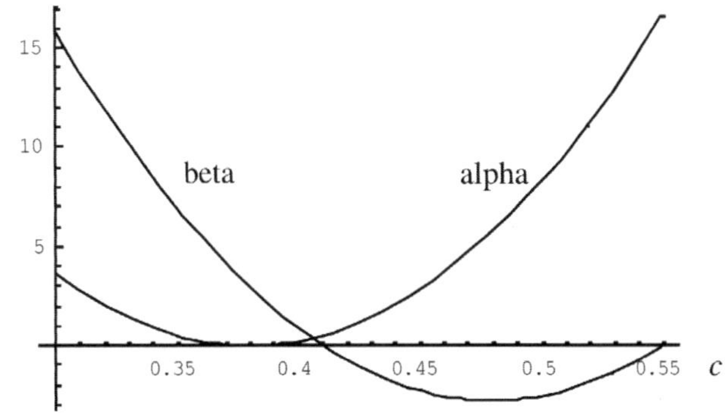

Figure 4. The chemical driving force for phase transformation.

The temporal evolution of the structural and the concentration fields are described by the time-dependent Ginzburg-Landau equation (1) and the non-linear Cahn-Hilliard diffusion equation (2) [1, 2], respectively,

$$\frac{\partial \eta_{\mathrm{p}}(\mathrm{r},t)}{\partial t} = -L\frac{\delta F}{\delta \eta_{\mathrm{p}}(\mathrm{r},t)} + \xi_p(\mathrm{r},t); \quad p = 1,2,...,6 \tag{1}$$

$$\frac{\partial c(\mathrm{r},t)}{\partial t} = M\nabla^2 \frac{\delta F}{\delta c(\mathrm{r},t)} + \zeta(\mathrm{r},t) \tag{2}$$

where L and M are kinetic coefficients characterizing structural relaxation and chemical mobility respectively (assuming that L and M are constants). F is the total free energy of the system, $\xi_p(r, t)$ and $\zeta(r, t)$ are the random noise terms related to fluctuations in the structural order parameter and concentration, respectively.

The microstructure evolution process can be simulated with above variants selection, energy and governing equations formation. For example, the phase transformation of alpha to gamma has been investigated using the phase field method [6]. The detailed nucleation process is shown in figure 5. The nucleation and growth process is clearly seen from the simulation results. The twin nucleation is the main manner during alpha to gamma phase transformation. The twin frequency can be obtained with detailed statistical analysis, shown in figure 6. Furthermore, the influence of input parameters, such as the elastic energy coefficient, on the twin frequency can be obtained.

IOP Conf. Series: Materials Science and Engineering **422** (2018) 012011 doi:10.1088/1757-899X/422/1/012011

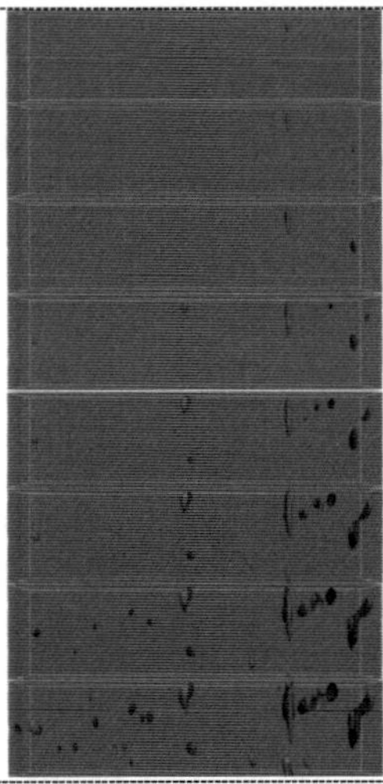

Figure 5. The nucleation process in the phase transformation of alpha to gamma.

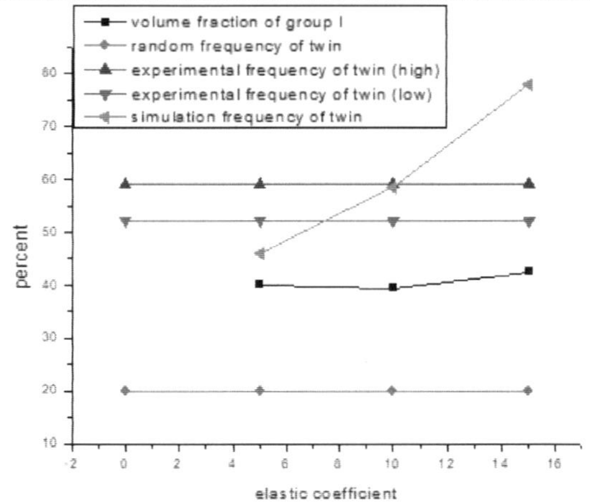

Figure 6. The statistical results of the nucleation process.

3. Discussions

The phase field simulation methods can be utilized in the research of solid phase transformation, liquid-solid phase transformation and the ferroelastic/ferroelectric/ferromagnetic phenomenon. Besides the phase transformation research in materials science and engineering, the application area of phase field simulation has been extended to biological research etc. in these years.

Though the phase field simulation method has success in the past years in the application of some research and industry sections. It is still in its very beginning stage. The existing phase field models can not satisfy the great demands of the materials design and other applications.

One problem is the lack of common and commercial phase field simulation software. The simulation program must be developed by the professional researchers. This difficulty in actual application restricts the wide and rapid spread of the phase field method.

Another problem is the difficulty in obtaining the input parameters of the phase field simulation. The accuracy of the phase field model depends on the accurate input parameters, such as the interfacial energy and elastic energy. Especially, the input parameters deviation with the change of temperature and composition, etc.

Therefore, more and more phase field models in different application aspects and more efficient algorithm should be developed in the upcoming period. Furthermore, the development trend of the computational materials science is integrating different simulation methods in different time scales and length scales.

It is a long way go and it is needed great effort on the development of the phase field simulation method before its easy and high efficient usage. Anyway, nothing can stand in the way of the phase field method becomes a powerful tool in microstructure evolution simulation.

4. Conclusions

The phase field method is a powerful tool in microstructure evolution simulation. The size, shape and spatial arrangement of phases, grains and so on can be obtained by this method. It will help the researchers tailor the microstructure of alloys. The relationship between the microstructure and the chemical driving force, the interfacial energy and the elastic strain energy can be explored in detail. With the development of computing power the phase field method will plays more and more important role in the microstructure design. Furthermore, the phase field method itself can be developed quickly.

5. References

[1] Wang Y, Chen LQ. Simulation of microstructural evolution using the field method. In: Kaufmann EN, editor. Methods in Materials Research. New York: John Wiley & Sons. Inc.; 2000. p.2a.3.1.
[2] Chen L Q 2002Ann Rev Mater Res 32 113.
[3] Wang Y Z, Li J 2010 Acta Mater 58 1212.
[4] Wen Y H, Chen L Q, Hazzledine P M and Wang Y 2001 Acta Mater 49 2341.
[5] Boettinger W J, Warren J A, Beckermann C, Karma A 2002 Ann Rev Mater Res 32 163.
[6] Teng C Y, Zhou N, Wang Y, Xu D S, Du A and Yang R 2012Acta Materialia 60 6372.

IOP Conf. Series: Materials Science and Engineering **422** (2018) 012012 doi:10.1088/1757-899X/422/1/012012

Fabrication of Porous Ceramics using Tailings Mud from Huzhou City as Raw Material

S W Yao[1], W B Zhang[1], Y F Yao[1], J G Yao[1], Y Y Wang[2] and H F Chen[2]

1 XinKaiYuan Crushed Stone Co. Ltd., Huzhou 313000, China
2 Department of Material Chemistry, Qiuzhen School, Huzhou University, Huzhou 313000, China

E-mail: arince@zjhu.edu.cn

Abstract. The comprehensive utilization of tailings mud has become a common concern of the society. The properties of tailings mud produced from Huzhou city were analyzed in this work. Moreover, porous ceramics with high strength were prepared by using the tailings mud as main raw material and adding a small amount of foaming agent. The results show that the foaming agent contents have a certain effect on the porosity and compressive strength of the porous ceramic samples. These results provided a certain theoretical basis for the comprehensive utilization of tailings mud and solve the pollution of tailings mud to the environment.

1. Introduction

Porous ceramic, which combines high specific surface area of porous materials with stability of ceramic materials, has become an important environmental material [1, 2]. The ability to control its pore size, porosity and bulk density has been improved with the development of technology [3, 4]. Compared with the traditional ceramic particle filter, there are many advantages for porous ceramic, such as energy saving, low cost, and good filtering effect. So, porous ceramics are widely used in the fields of water treatment, filter materials and automobile exhaust treatment [5].

There are many preparation methods of porous ceramics: adding pore forming agent, foaming method, organic precursor impregnation and sol-gel method [6, 7]. Besides, the preparation of raw materials is also diversified, such as chemical reagents, diatomite and kaolinite [8-10]. However, many researchers devote to develop different systems porous ceramics with solid wastes as raw materials in order to reduce the production cost and achieve sustainable utilization of resources [11-14]. Although the development of porous ceramics has made some progress, the research was mainly carried out in the laboratory due to the complexity of preparation technology of porous ceramics, and still far away from industrialization [15]. Moreover, the research on porous ceramics prepared by tailings mud is still in its infancy.

Tailings mud, as a common industrial solid waste, not only occupies a large amount of land resources, but also pollutes the environment seriously. The utilization of tailings mud has become a common concern of the society. In this work, we present the preparation of porous ceramics that could be used for water treatment and building materials. The porous ceramics of low volume density and high strength are prepared by foaming method with tail mud as raw material and SiC as foaming agent. The physical properties, crystal phase, and morphology of the porous ceramics were analysed. These results are meaningful to solve the problem of environmental pollution and the comprehensive utilization of the tailings mud.

2. Materials and methods

2.1. Preparation of porous ceramic samples
Porous ceramic samples were prepared via foaming process. Tailings mud, which was collected from Donglin Town, Huzhou City, was ground into a powder and mixed with a foaming agent and water. The samples were then pressed into strips with a tablet press, followed by further sintering in air by heating at a rate of 5 ºC/min. The preparation conditions of different samples were listed in Table 1.

Table 1. Preparation conditions of different samples

Sample	Sintering temperature (ºC)	Soaking time (min)	SiC (wt%)	Tailings mud (wt%)
T3	1130	30	3	97
T5	1170	20	1	99
T7	1200	15	2	98

2.2. Characterization
X-ray diffraction (XRD) analysis was performed with an XD-6 X-ray diffractometer at room temperature, with CuKα radiation, 36 kV voltage and 20 mA current. X-ray data were collected in the $10<2\theta<80$ range.
Differential thermal analysis (DTA) test was carried with a Thermogravimetric analyzer (CRY-2P, China). The sample mass was about 10.00 mg, and alumina was used as the reference. The data were collected in temperature range of 20~1100 °C in the air.
The particle size of the tailing mud was determined by Mastersizer 2000 laser particle size analyzer. Water was used as a dispersant in the process of experiment.
The porosity and volume density of the samples were determined using the vacuum method and the true density method [16].
Compressive strength was tested using an electronic universal testing machine (WDW-H10) following the standard test method for ceramic material compressive resistance (China, GB/T 4740-1999).

3. Results and discussions

3.1. Physical properties of tailings mud
XRD patterns obtained for the tailing mud samples are shown in Figure 1. Two major crystalline phases, i.e., quartz and albite, were observed in the samples. There is also some albite, Ca-rich in the sample. The individual contents of the different crystalline phases of the samples were also determined from the XRD patterns. The relative contents of quartz, albite and albite, Ca-rich were 45%, 40% and 15% (wt%), respectively.

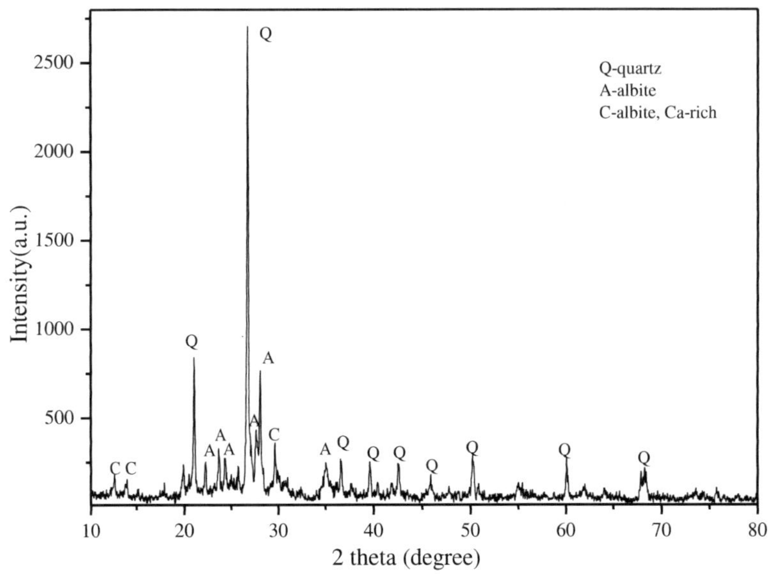

Figure 1. Room temperature XRD patterns of tailing mud

The size distribution of tailings mud was shown in Figure 2. It can be concluded that its D_{50}=14.00μm, D_{90}=51.66μm, the average volume particle size is 23.11μm while the specific surface area is 0.651 m^2/g. The size distribution width of particles is 3.408μm. The particle size is fine, the particle size distribution is uniform. and the reaction activity is good when used as the raw material.

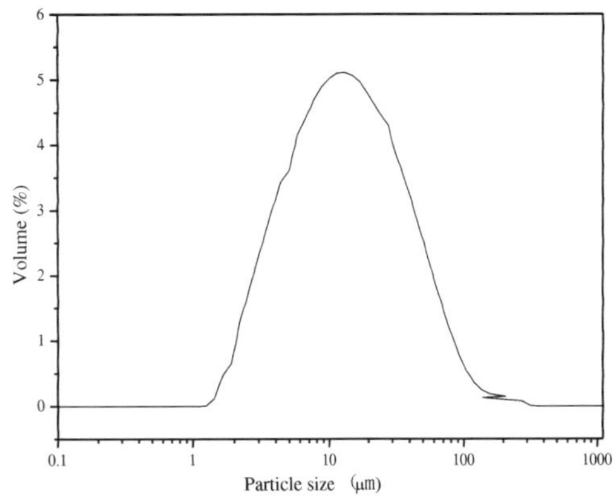

Figure 2. Particle size distribution of tailing mud

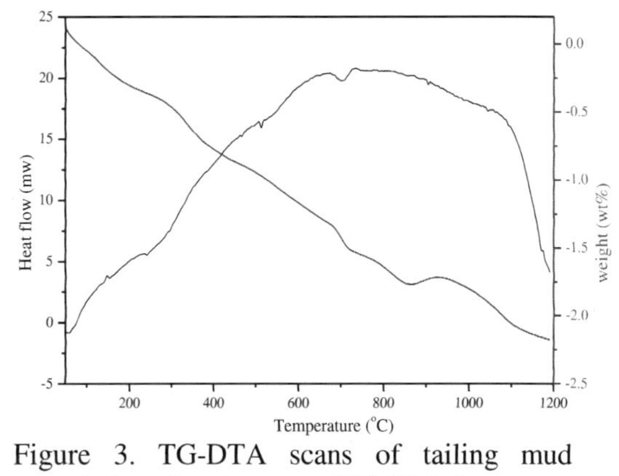

Figure 3. TG-DTA scans of tailing mud recorded at a heating rate of 20 °C /min

To illustrate the thermal behaviour of the tailing mud samples, a typical TG-DTA scan of tailing mud at a heating rate of 20 °C /min were shown in Fig. 3. Three weight loss stages were observed in the temperature ranges of 25-250 °C, 250-600 °C, and 600-1000 °C. The weight losses of about 0.4% at 25-250 °C could be ascribed to the evaporation of absorbed waters. The weight loss of 0.9% at 250-600 °C accompanied by a small endothermic peak might belong to the transformation of β-quartz to α-quartz. Finally, the endothermic peaks centred at 703 °C accompanied by the weight losses of 12.6% were likely due to the transformation of α-quartz to α-tridymite. It is generally believed that the transformation temperature of α-quartz to α-tridymite is 870 °C, which is a slow transformation. Because of the large difference in crystal structure between α-quartz and α-tridymite, the transformation energy is large, the transformation time is very long, and the transformation speed is very slow. However, because tailing mud samples contains Fe, Ca and other mineralizer, and the particle size of the tailing mud is small. All these make the transformation temperature of α-quartz to α-tridymite to a lower temperature. Furthermore, in combination with XRD results, the exothermic

peak at 903 °C and 1050 °C could be ascribed to the oxidation reaction of SiC [17].
Based on the above analysis, the tailings mud can be used as the raw material for the preparation of porous ceramics because of the primary conditions of the composition, phase and grain size of tailings mud.

3.2. Surface morphology and crystal phase of porous ceramic samples
The photos of surface morphology of porous ceramic samples were shown in Fig 4. The samples have smooth surfaces and compact structures. Both macroporous and mesoporous exist in all samples. However, the sizes of the pore are different and the distributions are uneven. The reason could be that the raw material and the foaming agent were not evenly stirred in the preparation process, which made the size distribution of the pore uneven after the sintering.

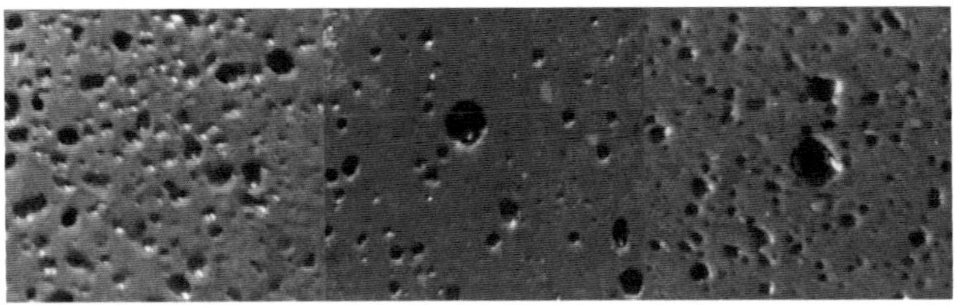

Figure 4. Photos of surface morphology of porous ceramic samples

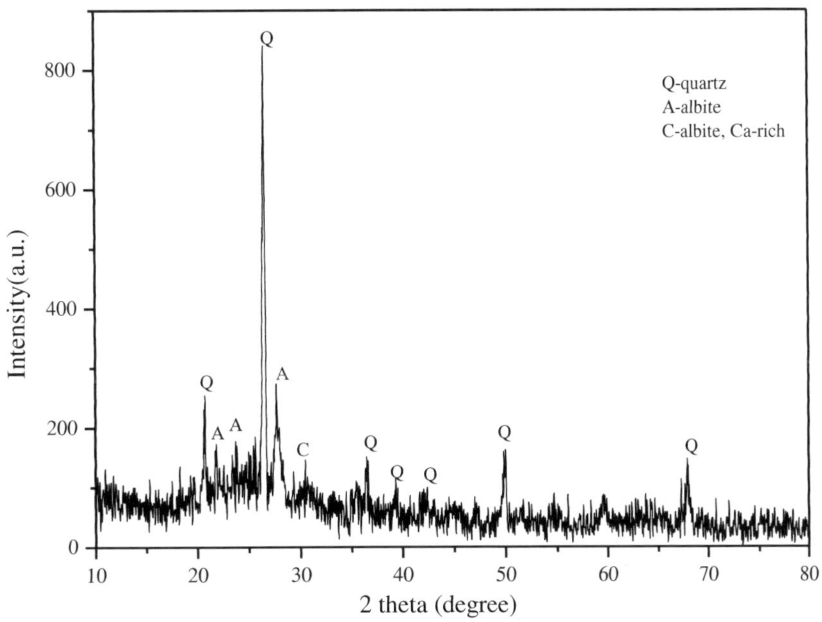

Figure 5. Room temperature XRD patterns of porous ceramic sample T3

XRD patterns of the porous ceramic sample T3 was shown in Fig. 5, demonstrating the presence of a mixture of crystalline phases. Two major crystalline phases, quartz and albite were observed in sample. In addition to those phases, albite, Ca-rich phase was also presented in the XRD pattern of sample T3. The phase of the porous ceramic sample is basically unchanged before and after the sintering compared to the raw materials. SiC was not observed in the XRD pattern of porous ceramic sample. During the sintering process, the SiO2 protective film formed on the surface of SiC react with feldspar melt to form silicate liquid phase, resulting in the destruction of the protective layer, so that

SiC was continuously reacted with oxygen to generate a large amount of CO or CO_2 gas, resulting in the disappearance of silicon carbide [18].

3.3. Physical properties of porous ceramic samples

Table 2. Physical properties of porous ceramic samples

Sample	Porosity (%)	Apparent porosity (%)	Volume density	Compressive strength (MPa)
T3	28.37	0.04	1.51	12
T5	29.24	0.33	1.74	22
T7	22.81	0.02	2.10	20

The physical properties of the porous ceramic samples, as determined by the vacuum method and an electronic universal testing machine, are listed in Table 2. Porosities of all the samples are over 20%, indicating that the samples are of moderate porosity. The foaming agent content has a certain influence on the porosity of the sample. The CO or CO_2 gas formed by foaming agent cannot be discharged rapidly in silicate liquid phase due to the liquid bond resistance. Therefore, a large number of pore structures are formed in the liquid phase, and the closed pores account for the vast majority of the pores. The compressive strength of the sample is up to 20MPA, which is related to the crystalline phases of the sample. The results of XRD show that the major phase in the sample are quartz, albite and other minerals. The formation of feldspar, quartz and other amorphous silicate materials and glass phases because of the melting and sintering of raw powder particles at high temperatures, which made the porous ceramic samples with higher strength and greater hardness.

4. Conclusions

In this work, porous ceramic samples were prepared via foaming method with tail mud as raw material. The results show that the tailings mud can be used as the raw material for the preparation of porous ceramics and foaming agent content impacts strength and pore size distribution of the resulting porous glass ceramics, which could provide a theoretical basis for the establishment of the preparation process of porous ceramic.

5. Acknowledgments

This research was financially supported by Natural Science Foundation of Zhejiang Province (LQ16E020001) and Science and Technology Planning Project of Huzhou City (2017KT04).

6. References

[1] Nettleship I 1996 *Key Eng. Mater.* **122** 305
[2] Hammel E C, Ighodaro L R and Okoli O I 2014 *Ceram. Int.* **40** 15351
[3] Ohji T and Fukushima M 2012 *Int. Mater. Rev.* **57** 115
[4] Liu R, Xu T and Wang C A 2016 *Ceram. Int.* **42** 2907
[5] Deville S 2017 *Adv. Eng. Mater.* **10** 155
[6] Han Y, Li J and Wei Q 2002 *Mater. Rev.* **16** 26
[7] Qian J M, Cui K and Hao A I 2005 *Ord. Mater. Sci. Eng.* **28** 60
[8] Zhang X, Liu X and Meng G 2005 *J Am. Ceram. Soc.* **88** 1826
[9] Yuan L, Yu J and Ye R 2013 *J Ceram. Process. Res.* **14** 525
[10] Yan W, Chen Q and Lin X 2015 *J. Ceram. Soc. Jpn.* **123** 897
[11] Ji R, Zhang Z, He Y, Liu L and Wang X 2015 Constr. Build. Mater. **85** 119
[12] Liu R, Xu T and Wang C A 2016 *Ceram. Int.* **42** 2907
[13] Hua K, Shui A, Xu L, Zhao K and Zhou Q 2016 *Ceram. Int.* **42** 6080
[14] Nishida T, Morimoto A, Yamamoto Y and Kubuki S 2017 *Appl. Water Sci.* **6** 1
[15] Guo D 2013 *China Ceram. Ind.* **20** 1

[16] Li B, Wang Y Y and Luo W Q 2107 *Mater. Sci. (Medžiagotyra)* **23** 356
[17] Badinia C, Finoa P and Ortonab A 2002 *J Eur. Ceram. Soc.* **22** 2071
[18] Shruti T W, Balasubramaniam R and Gupta M 2007 *Corros. Sci.* **49** 711

Study on Hybrid Effects of FRP Anchors for Strengthening of Concrete Structures

X. Wang[1a], J.W. Pan[1b], J.F. Xia[1], F. Wu[1]

[1]Nanjing University of Aeronautics and Astronautics, China

E-mail: [a]972881345@qq.com, [b]ken76@263.net

Abstract. FRP (Fiber Reinforced Polymers) has been widely used in structure strengthening with the advantages of high strength, light weight, corrosion resistance and so on. FRP anchor is an effective anchoring method for external strengthening FRP laminate. In order to reduce cost and improve performance, B/C HFRP (hybrid of basalt fiber/carbon fiber reinforced polymer) anchor was made through mixing the basalt fiber with conventional CFRP anchors. The shear behavior and the anchoring performance of two kinds of FRP anchors under different diameter and arrangement methods were tested by several groups of experiments. The experimental results shew that the ultimate strength, material utilization and ductility of specimens were improved in different degrees, which indicated that it was practicable in concrete structures strengthening with this hybrid method.

1. Introduction

In order to prevent or delay the debonding failure of FRP- concrete interface, and to improve the utilization ratio of materials and the effect of strengthening, the additional anchor of FRP sheets is particularly important. The commonly used anchoring methods are as follows: (1) mechanically-fastened fiber reinforced polymer (MF-FRP) [1], (2) U type hoop anchor [2], (3) near-surface mounted (NSM) FRP strengthening [3], (4) FRP anchor. Among them, FRP anchor is an emerging anchoring measure for external strengthening FRP laminate, which has aroused the concern of more and more scholars in recent years. J. G. Teng et al [4] tested on RC cantilever beams strengthened by GFRP sheets using FRP anchors. The result of experiments shew that FRP anchors could significantly improve the load capacity performance of strengthened beams, and reduce the FRP debonding speed, making full use of high-strength FRP strengthening material. H. W. Zhang et al [5] studied on the various factors which had impact on the fiber anchor. M. A. Alam et al [6] improved the FRP anchor system, using a new anchoring method which was a hybrid of multi-direction. I. S. Kim [7] applied the FRP anchor on strengthening for anti-progressive collapse. However, there are few studies on the hybrid FRP anchor, and the hybrid effects on the anchor performance remain to be further studied.

Considering the problem that the FRP anchor had relative low shear strength and the ductility of strengthened concrete structure is insufficient, B/C (basalt fiber/carbon fiber) HFRP anchor [8] was made through mixing the basalt fiber with conventional CFRP anchors. The destructive tests of a series of strengthened specimens were carried out by changing the diameter of the anchor fan, the root diameter of the FRP anchor and the arrangement of FRP anchors.

IOP Conf. Series: Materials Science and Engineering **422** (2018) 012013 doi:10.1088/1757-899X/422/1/012013

2. Experiment Study on the Performance of B/C HFRP Anchors

2.1. Preparation of B/C HFRP Anchor

In order to get relative high load carrying capacity and sufficient ductility, the approximate ratio of carbon fiber and basalt fiber was 1:2. According to the depth of pre-excavated hole (3cm depth) of concrete specimen and HFRP anchor fan diameter (1.5cm and 3cm), the length of two batches of FRP anchor was determined as 4.5cm and 6cm respectively; each bundle of basalt fiber and carbon fiber was tied with a string respectively; the diameters of the two batches of FRP anchor shall be 8 and 12 bundles of fibers; Then, a total of 8 or 12 bundles of basalt fibers and carbon fibers were tied together with strings to form a large mass of bundles.

2.2. Experimental Description

Simple shear test was conducted. The end of FRP laminate extending out of the concrete block was connected to the loading device for exerting a tensile force. To eliminate the stress concentration resulting from the uneven stress in the chuck, there must be sufficient distance between chuck and the test area. The specimens were divided into two categories, namely, control group layout (group A), single hole arrangement (group B) and four holes arrangement (group C). The specific layout of each group is shown in Table 1. The loading equipment of the test was shown in Fig. 1. In order to prevent the influence of the length error on both side, the connector and the force sensor was designed to hinged connection, which could keep the tension in the center of the FRP sheet.

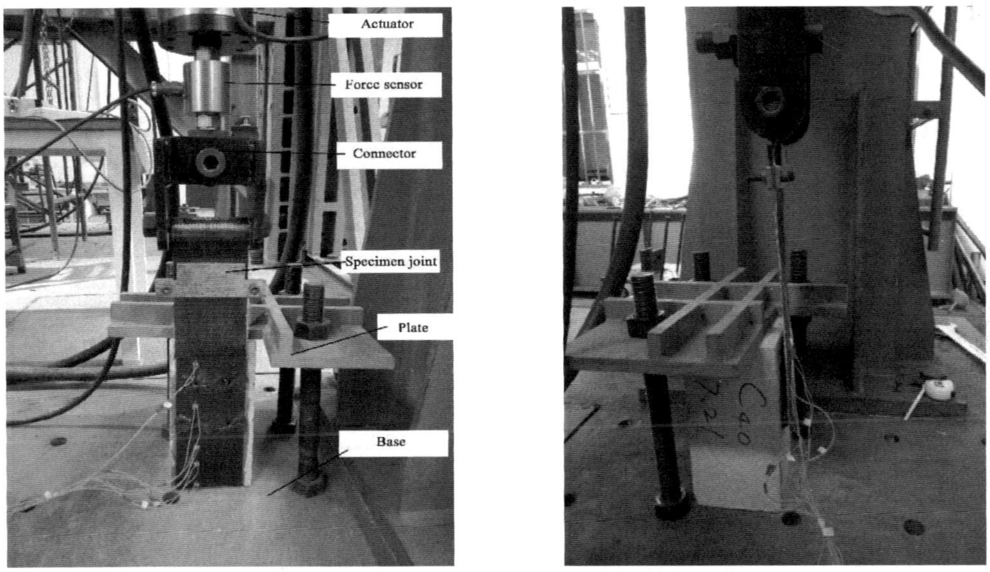

Figure 1. The loading equipment of the test

Table 1. Overview of the arrangement of each specimen

Specimen	The arrangement of FRP anchors	The diameter of FRP anchor fan [cm]	The root diameter of FRP anchors	The bond length of the FRP laminate [cm]
A-0-0-40	0	0	0	40
B-3-8-40	Single span	3	8 bundles	40
B-1.5-12-40	Single span	1.5	12 bundles	40
B-3-12-40	Single span	3	12 bundles	40
C-1.5-8-40	Four span	1.5	8 bundles	40
C-3-8-40	Four span	3	8 bundles	40
C-1.5-12-40	Four span	1.5	12 bundles	40

IOP Conf. Series: Materials Science and Engineering **422** (2018) 012013 doi:10.1088/1757-899X/422/1/012013

3. The Experimental Results

(1) For A-0-0-40 (Fig. 2a), its failure mode was the debonding of the bond end of the FRP sheet, then the specimen lost carrying capacity. Its ultimate load carrying capacity was 10.2KN.
(2) For B-3-8-40 (Fig. 2b), its failure started on the root of FRP anchors, then the FRP sheet and the anchors debonded together. Its ultimate load carrying capacity was 13.2KN.
(3) For B-1.5-12-40 (Fig. 2c), its failure started on part of FRP anchors fan torn, then the overall debonding happened in the surrounding FRP sheet of FRP anchors, but the root of FRP anchors did not break. It's ultimate load carrying capacity is 14.4KN.
(4) For B-3-12-40 (Fig. 2d), the root of FRP anchors fractured, then the FRP sheet had the overall debonding. Its ultimate load carrying capacity was 15.4KN.
(5) For C-1.5-8-40, C-3-8-40, C-1.5-12-40, they all have the same destructive characteristics (Fig. 2e). The debonding area was extended to the rear of the two bottom FRP anchors, and the FRP fiber sheet around the four FRP anchors was completely peeled off. The ultimate load carrying capacity of C-1.5-8-40 was 19.2KN. The ultimate load carrying capacity of C-3-8-40 was 18.6KN. The ultimate load carrying capacity of C-1.5-12-40 was 19.6KN.
By compared between the specimens with FRP anchors and the control specimens, we could find that (Fig. 3, Fig. 4) all the ultimate load carrying capacity of the specimens with FRP anchors were raised in different degrees, and the anchoring effect was fine, so the utilization ratio of the FRP sheet was improved.

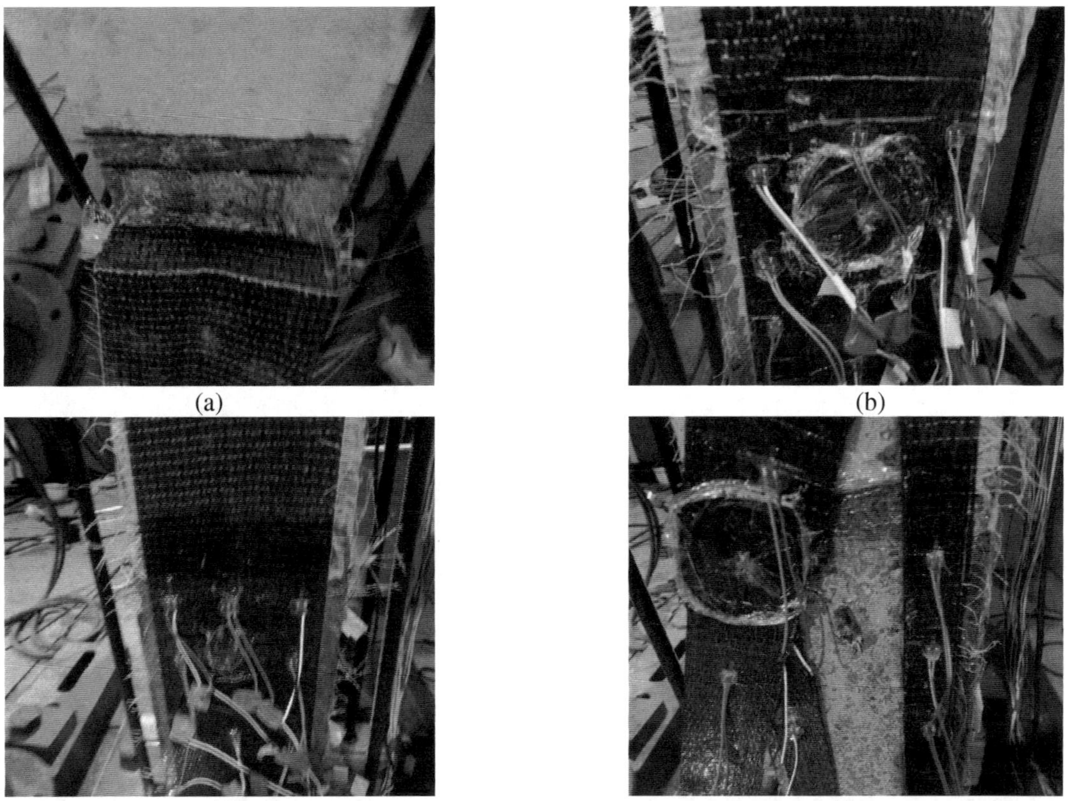

(a) (b)

(c) (d)

(e)

Figure 2. The failure modes of (a) A-0-0-40; (b) B-3-8-40; (c) B-1.5-12-40; (d) B-3-12-40; (e) Four Holes Arrangements

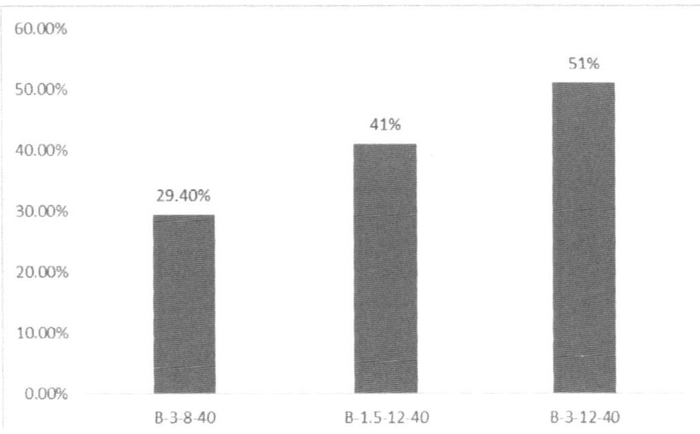

Figure 3. The performance improvement percentage of ultimate load carrying capacity of Group B

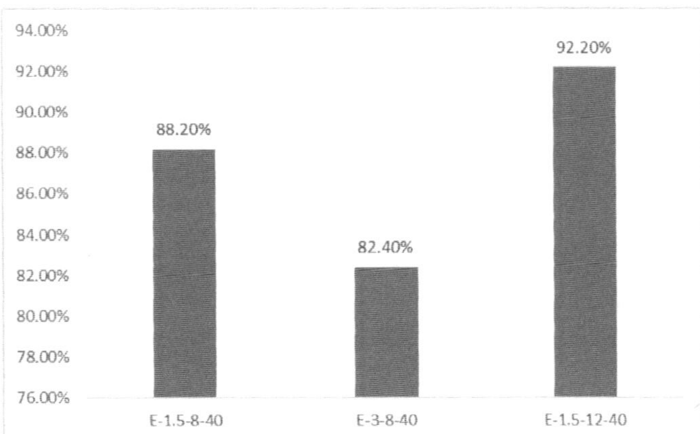

Figure 4. The performance improvement percentage of ultimate load carrying capacity of Group C

4. Conclusion

(1) Using the B/C hybrid anchors whose hybrid ratio of carbon fiber and basalt fiber was 1:2 could effectively improve the utilization of fiber fabric. At the same time, the addition of cheaper basalt fiber could greatly reduce the cost.

(2) By comparing the ultimate load capacity of the three members of E group, it could be found that the ultimate load capacity of each member is not much different and they have the same failure mode whose failure did not occur in the FRP bonding area, which indicated that the stronger constraints placed at the end of the FRP bonding area could not increase the ultimate load capacity unlimitedly.

(3) The placement of multi-pass FRP anchors on the FRP sheet could transfer the stress to the back end of the FRP bonding area, so that more FRP sheet could be involved in the force bearing, and the longitudinal utilization ratio of FRP sheet could be increased, and the overall ultimate load capacity of the component could be greatly improved.

5. Acknowledgment

The authors are grateful for the support provided for this research by the National Key Research and Development Program of China (No. 2016YFC0701400), and the Fundamental Research Funds for the Central Universities (No. NS2016018).

6. Reference

[1] L.C. Bank.: submitted to FRP Composites in Civil Engineering (2004).

[2] J.B. Quattlebaum, K.A. Harries, M.F. Petrou. *Comparison of Three Flexural Retrofit Systems under Monotonic and Fatigue Loads*. Journal of Bridge Engineering. Vol. 10(2005), p. 731-740.

[3] Rizzo, Andrea, L.D. Lorenzis. *Behavior and capacity of RC beams strengthened in shear with NSM FRP reinforcement*. Construction & Building Materials. Vol.23 (2009), p. 1555-1567.

[4] J.G. Teng, L. Lam, W. Chan, et al: submitted to Journal of Composites for Construction (2000)

[5] H.W. Zhang, S.T. Smith: *Influence of FRP anchor fan configuration and dowel angle on anchoring FRP plates*. Composites Part B: Engineering.Vol.43(2012), p. 3516-3527.

[6] M.A. Alam, M.T.E. Amin, M.Z. Jumaat: submitted to Indian Journal of Science and Technology(2012)

[7] I.S. Kim, J.O. Jirsa, O. Bayrak: submitted to Aci Structural Journal (2013).

[8] J.F. Xia: *Experimental Study on the performance of B/C HFRP Anchors*. Nanjing university of aeronautics and astronautics(2015).

Effect of Fibre Content on Compressive Strength of Wheat Straw Reinforced Concrete for Pavement Applications

Muhammad Usman Farooqi[1, a], Majid Ali[1, b]

[1]Civil Engineering Department, Capital University of Science and Technology, Islamabad, Pakistan

E-mail: [a]u.farooqi168@gmail.com, [b]drmajidali15@gmail.com

Abstract. Many studies have been carried out on natural fiber reinforced concrete for the different civil engineering applications. Wheat Straw Reinforced Concrete (WSRC) is recently studied for pavement applications which resulted in improved toughness. This will ultimately help in reducing the micro-shrinkage and fatigue cracking in rigid pavements. The overall aim of the research program is to improve the rigid pavement construction techniques by using locally available natural fibres. In this work, compressive strength of WSRC is investigated experimentally by varying fiber contents i.e. 1%, 2% and 3%, by mass of wet concrete. The properties of WSRC are compared with that of Plain Concrete (PC). The compressive behavior, strength, energies absorbed and toughness indices of PC and WSRC are determined and discussed. It is observed that, under the compressive loading, the fragments of PC specimen are chipped off while in WSRC specimen only cracks are formed. It is concluded that compressive toughness index of WSRC is 71% more than that of PC. Based on the conclusions made, WSRC shows favorable results, therefore, future recommendations are to focus on optimization and durability of WSRC.

1. Introduction

Concrete pavements have longer structural life and superior durability as compared to asphalt pavements due to which their use in developing countries has been grown over the past decade [1]. However, its more extensive use is still prevented due to its higher initial cost, mainly due to cost of cement and reinforcement. Moreover, plain concrete (PC) is inherently a brittle natured material having low strain capacity and tensile strength. Therefore, the fatigue failure can potentially be exhibited at more locations in concrete pavements due to the combination of built-in temperature curl, moisture gradients, temperature gradients, and drying shrinkage, as pavements are directly exposed to environmental conditions [2]. It has been long recognized that energy absorption capacity of PC can be enhanced by the addition of fibres [3, 4].

Fibres have been used to strengthen the brittle matrices since Biblical times; for example, horsehair and straw was mixed with clay to form bricks and floors. For the first time in modern technology, Romualdi [5] proposed the steel fibres as dispersed reinforcement for concrete. Effect of artificial fibres in concrete, for rigid pavements, has been investigated by many researchers [6] which resulted in improved properties of concrete.

Considerable efforts have been made for using different natural fibres in last few decades. As natural/vegetable fibres are cheap and available in abundance in tropical and sub-tropical regions.

Hence, it costs a very little as compared to the artificial/steel fibres. Natural Fibre Reinforced Concrete is comparable with Artificial Fibre Reinforced Concrete (AFRC) and Steel Fibre Reinforced Concrete (SFRC) for various civil engineering applications [7].
Natural fibres have already been investigated for concrete pavement applications with the limited scope [8, 9]. Results showed that natural fibres have the potential to be used as reinforcement in rigid pavements. However, the durability of natural fibres is still questionable till today. So, the application of natural fibres in the construction industry is still quite limited [10]. The overall aim of the research program is the development of economic and durable design and/or construction techniques for new rigid pavements by using locally available natural fibres in concrete. Wheat straw has been selected to start with. As a part of ongoing research project, the effect of fibre content on compressive strength of WSRC is intended for the pavement applications.

2. Experimental Methodology

2.1 Raw materials
Ordinary Portland cement of a local company, fine and coarse aggregates from the Margalla crushes in Pakistan, and portable water were used for preparing PC. Wheat straw were extracted from the agricultural residues of wheat crop in the surrounding region during the time of wheat cultivation. The locally available wheat straw were prepared after dipping in water for 15–20 minutes. These prepared straw were named as natural wheat straw.

2.2 Mix proportion and casting methodology
The mix proportion of PC and WSRC was 1:2:4 (cement: sand: aggregate) with a water – cement ratio of 0.55 and 0.60, respectively. Straw having varying lengths of approximately one inch and contents of 1%, 2% and 3%, by mass of wet concrete, were considered to make Wheat Straw Reinforced Concrete (WSRC). For preparing PC, all materials were put in the drum mixer along with water. The mixer was rotated for a period of five minutes. At that stage, the homogenous and workable PC mix was formed. All PC mix was taken out of the mixer. Whereas, for WSRC, half of PC mix was taken and weighed and then, put it again in the mixer. Straw (as per considered content) were added in the mixer and the mixer was again rotated for one-minute period. The mix was not seemed to be homogenous and workable at that stage. A small increment of 0.05% in the water-cement ratio (i.e. 0.55% initially) was made. Then, the mixer was started again to rotate for a period of three minutes to have the better mix. The mixing time was increased because at that stage, the further addition of water could have been resulted in bleeding of WSRC. Therefore, the increased water – cement ratio by only 0.05% and increased mixing time of 4 minutes came out to be a successful approach for a workable WSRC. It may be noted that it was relatively easy to fill moulds with WSRC having 1% content compared to WSRC having 2% and 3% contents. However, moulds were filled with WSRC having 2% and 3% contents with little difficulty i.e. lifting and dropping were made relatively more.
The mix was poured into the moulds in three successive layers for preparing WSRC specimen. After having compaction of each layer by 25 blows of temping rod, the mould was lifted up to approximately 100-150 mm and dropped freely on the floor for removing air voids from WSRC mix and for self-compaction. All specimen were de-moulded after 24 hours and were kept for curing of 28 days in water tank before testing.

2.3 Specimen
PC and WSRC cylinders having dimensions of 100 mm diameter and 200 mm height were prepared for compressive strength tests. Labels PCC, A1, A2 and A3 were used for PC, WSRC with 1% content, WSRC with 2% content and WSRC with 3% content, respectively. Alphabets along with labels show the number of sample for each specimen. A total of 8 cylinders were cast. Average of two readings was taken for every property.

2.4 Testing methods

ASTM C39 standard was used to perform the compressive strength test on PC and WSRC specimen. Cylinder test specimen were tested after 28 days of curing to determine the compressive behavior, compressive energies absorbed, compressive energies absorbed up to the maximum load, compressive energies absorbed post the maximum load and compressive toughness indices of PC and WSRC.

3. Results and Analysis

3.1 Compressive behavior

Figure. 1. shows the load-time curves for PC, WSRC-1%, WSRC-2% and WSRC-3%. The noticeable thing is that the ascending side of WSRC-1% curve is comparatively less steep than that of PC. However, the WSRC cylinders bear the load over more time period. In spite of the fact that the curves for WSRC-2% and WSRC-3% are more flattened and bear less load, but even then, these are showing the tough behavior due to the presence of straw. The tested cylinder specimen of PC, WSRC-1%, WSRC-2%and WSRC-3% are shown in Figure 2. The fragments of PC specimen are broken, whereas, only cracks are observed in WSRC specimen. This shows the sewing effect and tough nature of WSRC due to the presence of straw.

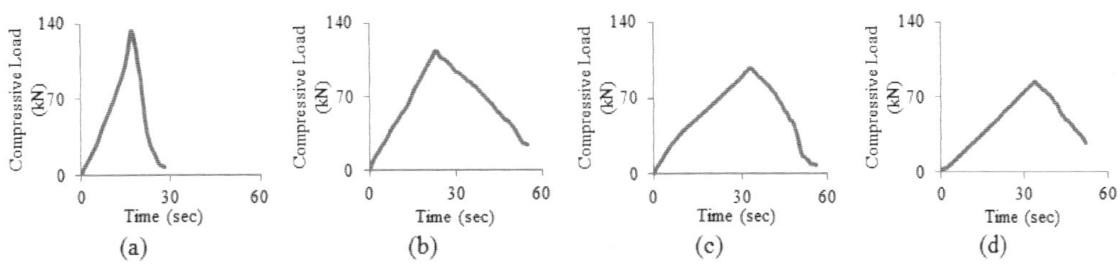

Figure 1. Compressive load – time curves for (a) PC, (b) WSRC-1%, (c) WSRC-2%, and (d) WSRC-3%

(a) (b) (c) (d)

Figure 2. Tested specimen of (a) PC, (b) WSRC-1%, (c) WSRC-2%, and (d) WSRC-3% under compressive loading

3.2 Compressive strength, compressive energies absorbed and compressive toughness index

Compressive strength (σ) is taken from the maximum load of load – time curve of Figure 1. The area under load – time curve up to the maximum load is taken as the compressive energy absorbed up to the maximum load (CEM). The area under load – time curve from the maximum load up to the failure is taken as the compressive energy absorbed post the maximum load (CEP). The total area under load – time curve is calculated for measuring the compressive energy absorption (CE). The compressive toughness index (CTI) is ratio of compressive energy (CE) to the compressive energy absorption up to the maximum load (i.e. CE/CEM). σ, CEM, CEP, CE, and CTI of PC, WSRC-1%, WSRC-2% and WSRC-3% are shown in Table 1. The σ of WSRC-1%, WSRC-2% and WSRC-3% are decreased by 14%, 27% and 37%, respectively, when compared to the σ of PC. Although, the σ of WSRC specimen is decreased, but the CE of WSRC-1%, WSRC-2% and WSRC-3% are increased by 136%, 103%, and 64%, respectively, which shows the tough behavior due to the presence of dispersed straw. Here, it can be observed that almost 63% of total energy is absorbed by WSRC-1% after the maximum load, whereas, in case of PC, only 38% of the total energy is absorbed after the maximum load. Furthermore, due to this sewing effect, an increase of 71%, 7% and 9% in CTI of WSRC-1%, WSRC-2%, and WSRC-3% is observed as compared to that of PC.

Table 1. Compressive strength, compressive energies absorbed and compressive toughness index of PC, WSRC-1%, WSRC-2%, and WSRC-3%

Specimen	б	CEM	CEP	CE	CTI
	[MPa]	[kN.s]	[kN.s]	[kN.s]	-
PC	2.06	937	575	1512	1.61
WSRC-1%	1.77	1296	2272	3568	2.75
WSRC-2%	1.50	1771	1293	3064	1.73
WSRC-3%	1.30	1408	1070	2478	1.76

The comparison of б, CEM, CE, and CTI of PC, WSRC-1%, WSRC-2% and WSRC-3% are shown in Figure 3. In spite of the fact that the б of WSRC-1% is decreased up to some extent but as a whole, WSRC-1% behaves well as compared to all tested matrices.

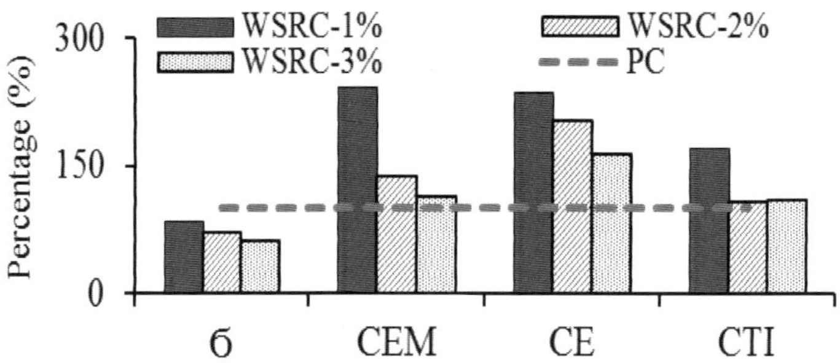

Figure 3. Comparison of б, CEM, CE, and CTI of PC, WSRC-1%, WSRC-2% and WSRC-3%

4. Discussions

The cracks are usually observed in pavements at very early stage. These cracks are more likely to propagate as pavements endure dynamic loading. Previous studies show that, with addition of fibres, the energy absorption capacity of concrete can be enhanced [4]. Hence, the straw's sewing effect in the concrete can resist the crack propagation, ultimately increasing its energy absorption capacity.

The tested matrices with 2% and 3% straw content show a slight improved behavior in energies absorption and toughness indices when compared to PC but at the same time, a significant decrease in compressive strength is observed. This might be due to the less compaction or improper mixing of the matrices due to the more content. Among tested matrices, WSRC-1% behaves well in terms of improved CEM, CEP, CE and CTI in spite of comparable б (even on lower side) in comparison with that of PC. So, it is likely that rigid pavements with WSRC-1% would behave better than pavements with PC.

5. Conclusions

The compressive behaviors of plain concrete, wheat straw reinforced concrete with 1% content, wheat straw reinforced concrete with 2% content, and wheat straw reinforced concrete with 3% content are investigated for possible applications in rigid pavements. Following conclusions are made on the basis of conducted study:

- Compressive strengths of WSRC-1%, WSRC-2% and WSRC-3% are decreased by 14%, 27% and 37%, respectively, when compared to the compressive strength of PC.
- An increase of 136%, 103%, and 64% is observed in the total energies absorbed by WSRC-1%, WSRC-2% and WSRC-3%, respectively, when compared to the energy absorbed by PC.

- CTI of WSRC-1%, WSRC-2% and WSRC-3% is increased by 71%, 7% and 9%, when compared to the CTI of PC. WSRC-1% shows the bridged behavior, whereas, PC shows brittle behavior.

So, it can be claimed that wheat straw reinforced concrete has the potential to be used for pavement applications due to its better post-cracking behavior. WSRC-1% behaves very well among all other tested matrices. Being an initiative study, the results discussed here are based on limited testing and analysis, the outcome seems favorable which demands for a detailed investigation focusing on optimization of materials and durability.

6. Acknowledgment

The authors would like to acknowledge all the organizations and persons who helped them during the research, particularly Mr. Muhammad Khalid and Mr. Muhammad Junaid for their kind help during lab work. The authors would also like to gratefully acknowledge the anonymous reviewers for the careful review and constructive suggestions.

7. References

[1] J.E. Hiller and J.R. Roesler: Journal of Transportation Engineering, 131(8), 2015.
[2] K. Smith, M. Wade and D. Peshkin: FHWA-RD-95-110, 1996, Washington, D.C.
[3] H.W. Reinhardt and A.E. Naaman: 3rd Intl. RILEM Workshop, 1999.
[4] C. Achilleos, D. Hadjimitis, K. Neocleous and S. Kallis: Sustainability, 3, 2011.
[5] J.P. Romualdi and G.B. Batson: Journal of Engineering Mechanics, 89, 1963.
[6] N. Salemi and K. Behfarnia: Construction and Building Materials, 2013, 934-941.
[7] M. Ali and N. Chouw: Construction and Building Materials, 30, 2012, 814-825.
[8] D. Patel and V.M. Patel: Indian Journal of Research, 4(4), 2015, 4-5.
[9] U. Farooqi and M. Ali: SCMT – 4, August 7-11, 2016, Las Vegas, USA.
[10] F.A. Silva and J.A. Filho: High FRC Composites, 2007, 175-182.

IOP Conf. Series: Materials Science and Engineering **422** (2018) 012015 doi:10.1088/1757-899X/422/1/012015

Study on the Characteristics of Electrothermal Corrosion Trace of Aluminium-Brass Connector

Dazhi E

Shenyang Fire Research Institute, Shenyang, China

E-mail: raul9055@163.com

Abstract. In this paper, both thermal corrosion trace and electrothermal corrosion trace of aluminium-brass connectors are prepared by conducting simulation experiments under different heating temperatures and overload current conditions. The test samples are analyzed and described by metallographic analysis method, and the corresponding characteristic patterns are summarized, which provides the technical basis for the identification of the melting traces in the fire.

1. Introduction

Fire investigation is the responsibility of the public fire protection institutions in accordance with Fire Law of People's Republic of China, is a kind of systematic, rigorous and scientific work. A scientific, accurate conclusion on the cause of the fire is supported by a lot of evidences, which must have no relationship with each other, but supported by each other to be a complete “chain of evidence”. Therefore, identification the material evidences of the fire scene accurately is a process of forming "chain of evidence" with independent evidences, which is the core and technical guarantee in the determination of the cause of the fire. It is an important component of investigation on electrical fire to accurately extract the electrical fire evidence and identify it, and even determine whether the cause of the fire is correct or not. For material evidence identification technology of electrical fire, in the persistent efforts of the researchers in our country, the application of macro analysis, metallographic analysis, morphology and composition analysis and other technical methods, have basically solved qualitative and semi-quantitative evaluation problem of the material evidence of electrical fire. However, there are some identification problems of special traces have not been resolved, such as identification problem of melting trace for electrical connector. Due to the connection of the plug-in often employs the different metals such as aluminium conductor connected to brass connectors, while the contact resistance and the membrane resistance between different metallic materials are both high, in the course of long-term use or overload use, it tends to form metal melting trace on the connecting point [1-3]; On the other hand, when the different metals are subjected to thermal effect, the diffusion free energy of the materials can be reduced, which often results in the interfusion of the conductor and the connector below the melting point of either metal. As lack of related research, it is a technical problem to identify and distinguish between the thermal corrosion trace and electrothermal corrosion trace of electrical connector, which brings the trouble to determine the cause of the fire in some special circumstance. Therefore, we need to carry out the related research, in order to solve the problem of existing methods and correspond to the requirements of fire investigation.

2. Simulation test and sample preparation

For studying the effect of thermal corrosion and electrothermal corrosion on electrical component under the circumstance of high temperature of fire and electrical failure, in this paper, solid aluminium conductor (2.5) and brass connector (10 A) are selected as the main research object which are common material evidences in fire.

Figure 1. Appearance of connector at the heating temperature of 500 ℃

Figure 2. Appearance of connector at the heating temperature of 600 ℃

2.1. Heating test of connector

To simulate the lean oxygen environment in actual fire scene, place connectors at high temperature and vacuum resistance furnace (rate of temperature rising with 800 °C/hour) for heating, setting up test temperature at 400 °C, 500 °C, 600 °C, 700 °C and 800 °C, respectively. When reaching the required test temperature, keep the temperature stable for 30 minutes, then cooling the sample with the furnace naturally(cooling rate is 200°C/hour). After the tests, the samples are analyzed accordingly.

The appearance of the connector after the test is as follows:

(1) At the heating temperature of 400°C and 500°C, the connector has no significant change, as shown in Figure 1.

(2) At the heating temperature of 600°C, the connector has no significant change, on which the connected solid aluminium conductor is slightly fused, as shown in Figure 2.

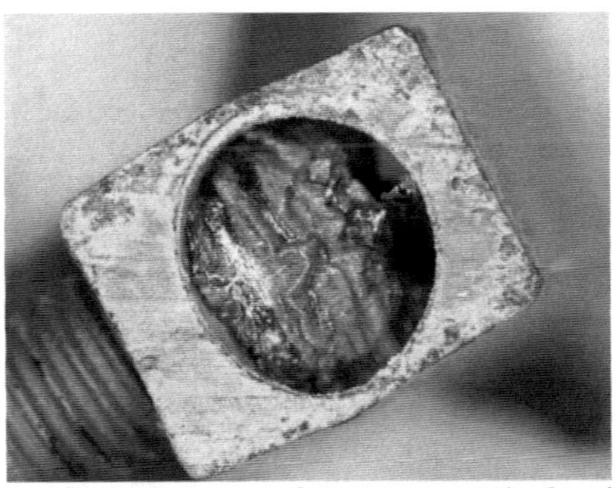

Figure 3. Appearance of connector at the heating temperature of 700 ℃

Figure 4. Appearance of connector at the heating temperature of 800 ℃

(3) At the heating temperature of 700°C, the connector appears slight deformation, solid aluminium conductor and brass connector are melt slightly, as shown in Figure 3.

(4) At the heating temperature of 800°C, the deformation of connector is serious, solid aluminum conductor and brass connector are mutually melt together, as shown in Figure 4.

2.2. Overcurrent test of connector

Connectors are joined up with electrical load under the condition of AC220V. The overcurrent experiments are carried out at the rated current of 2.0Ie, 3.0Ie, 4.0Ie and 5.0Ie (20-group tests are conducted for each overcurrent condition), the connection mode is shown as Figure 5. During the test, the temperature parameters near the connectors are recorded by infrared thermal imager and thermocouples, then after the test, the test samples are analyzed accordingly.

Figure 5. Overcurrent test of connector

Figure 6. Overcurrent test of connector at 3.0Ie

The detailed phenomena of tests are as follows:

(1) At the rated current of 2.0Ie, the temperature at contact point is going up slowly. And the equilibrium temperature is usually achieved in 2-3 minutes, which is to maintain under 110°C. There is no obvious change on both the connector and the insulation of conductor.

(2) At the rated current of 3.0Ie, the temperature at contact point is rising fast. And the equilibrium temperature is usually achieved in about 2 minutes, which is to maintain at 200°C. It appears a slight thermal decomposition on the insulation layer of conductor, as shown in Figure 6.

Figure 7. Overcurrent test of connector at 4.0Ie

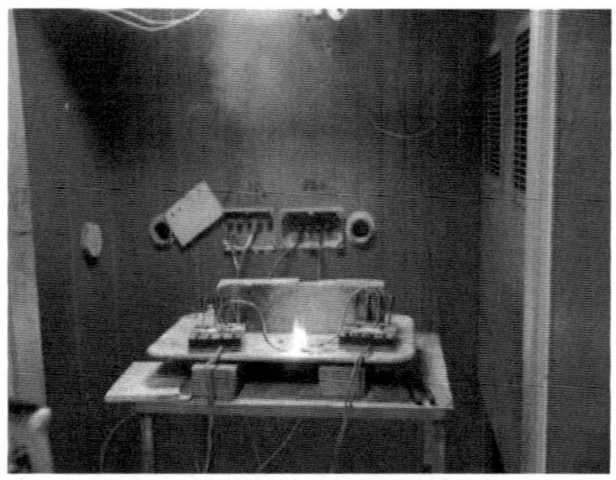

Figure 8. Overcurrent test of connector at 5.0Ie

Figure 9. Overcurrent trace of connector at 4.0Ie

Figure 10. Overcurrent trace of connector at 5.0Ie

(3) At the rated current of 4.0Ie, the temperature at contact point is rising rapidly. It appears a significant thermal expansion on the insulation layer of conductor and a large amount of smoke are produced, as shown in Figure 7.

(4) At the rated current of 5.0Ie, the temperature at contact point is rising rapidly, some samples have the phenomenon of melting and arcing, and flame combustion is observed, as shown in Figure 8. After the tests, the connectors are removed for macroscopic examination, some of which appear discoloration, and some of which have the situations of breakdown or melting traces, as shown in Figure 9 and Figure 10

3. Sample analysis and pattern summary

By means of metallographic analysis, composition analysis and other technical methods, the samples prepared by simulation tests are analyzed and the corresponding characteristic patterns are summarized.

3.1. Heating sample of connector

The boundary between the aluminium conductor and the brass connector is obvious, which are the multi-layer structure (the two layers). The original body of aluminium conductor is characterized by large columnar crystals or isometric crystals, while the original body of brass connector appears obvious annealing. The higher the temperature, the greater the thickness of the diffusion layer. Metallographic photos are shown in Figure 11.

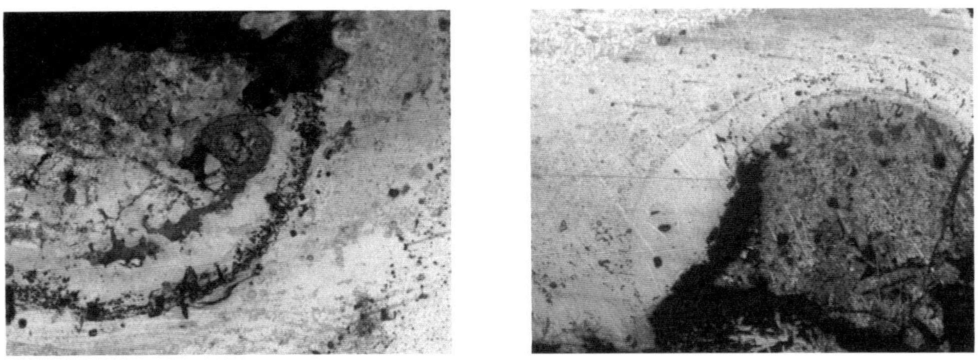

(a) (b)

Figure 11. Metallographic structures of heating sample of solid aluminium-brass connector 100X

According to composition analysis, there are two main elements of copper and aluminium in the diffusion layer, and there are also obvious differences in the contents of two elements in the diffusion layer, that is, the content of copper element near the original body of brass connector is obviously higher.

3.2. Overcurrent sample of connector

It presents a layered structure on the metallographic structure, but only one of which is transition layer. The side of solid aluminium presents tiny columnar crystals and acicular crystals with obvious direction [4, 5]. The side of brass connector possesses certain heat-affected zone, as shown in Figure 12.

(a) (b)

Figure 12. Metallographic structures of overcurrent sample of solid aluminium-brass connector 100X

3.3. Comparative analysis of sample characteristics

The comparison of the characteristics of the above sample structures can be seen:

(1) There is a certain mutual melting phenomenon while either electrothermal corrosion or thermal corrosion is having effect on the electrical component with different types of materials.

(2) The electrothermal corrosion traces formed by the arc discharge are similar to the traces formed by traditional ground fault.

(3) At the heating temperature of 600°C, although aluminum conductor has contracted with brass connector, there is not thermal corrosion happened.

(4) Although it presents a layered structure on both electrothermal corrosion and thermal corrosion, there are obvious differences in the number and morphology of the diffusion layer, which can be considered as an important condition for deciding trace type.

4. Conclusions

In this paper, both thermal corrosion trace and electrothermal corrosion trace of aluminium-brass connectors are prepared, testing phenomenon and the macroscopic characteristics of test samples are described. On the basis of metallographic analysis and composition analysis, characteristic patterns of traces are summarized, which will provide the technical support for the identification of the melting traces in the fire.

5. References

[1] Ministry of Public Security Fire Station 2013 *China Fire Service 2013* (Beijing: International Culture Press)

[2] Ministry of Public Security Fire Station 2014 *China Fire Service 2014* (Beijing: International Culture Press)

[3] Ministry of Public Security Fire Station 2015 *China Fire Service 2015* (Beijing: International Culture Press)

[4] Jinzhuan Zhang and Hao Jiang 2008 *J. Fire Safety Science.* **17(1)** 63-66

[5] Shanjun Mo, Wenjing Peng, Dong Liang and Yutao Long 2012. *J. Journal of Safety Science and Technology.* **8(1)** 63-70

Digital Image Correlation (DIC) Technique for Fracture Toughness Calculation of Microalloyed Steel (38MnVS6)

Akshay B. Patil[1, a], Suraj P. Toppo[1], R K P Singh[1]

[1]Kalyani Centre for Technology & Innovation, Bharat Forge Ltd., Pune- 411036, Maharashtra, India.

E-mail: [a]akshay.patil@bharatforge.com

Abstract. The present work shows the practical application of digital image correlation (DIC) technique for fracture toughness (K_{IC}) calculation. DIC is a non-contact, optical technique for measuring full-field displacements and strains by comparing an image of a deformed specimen surface to a reference image taken at un-deformed state. These captured images are correlated and the surface displacements are calculated from which the crack mouth opening displacements (CMOD) is determined. Results of DIC fracture toughness (K_{IC}) value is validated by comparison with fracture toughness (K_{IC}) value calculated from clip gauge. It is observed that the results obtained by DIC are in close agreement with the results obtained through conventional methods.

1. Introduction

The main applications of microalloyed steel (38MnVS6) in automobile industries are for manufacturing engine crankshaft, connecting rod, front axle beam, steering knuckle etc. [1]. Microalloyed steel (38MnVS6) forged part are cost effective for manufacturing engine crankshaft in higher production volume. Therefore most of the crankshafts are made by using microalloyed steel (38MnVS6). The Crankshaft undergoes repeated bending & shear stresses during its working condition. It causes fatigue failure of the crankshaft by creating the cracks in fillet and/or oil hole area where strain is most severe [2]. Fatigue failure is the main failure mechanism for crankshaft subjected to dynamic loadings. In general, fatigue failure consists of three stages namely, crack initiation, crack propagation and final fracture. Thus, the whole lifecycle of the any dynamically loaded crankshaft is the loading cycles for crack initiation and crack propagation. The loading cycles of crack initiation depend on crack initiation resistance of material being used. The crack propagation resistance is measured through a material property called fracture toughness (K_{IC}). The fracture toughness (K_{IC}) values may serve as a basis in material characterization, performance evaluation, and quality assurance for typical engineering structures, including nuclear pressure vessels, automotive, ship and aircraft structures etc. [3].

There are numerous techniques are used to determine the fracture mechanics parameter but for accurate determination still required several challenges. This includes the accurate determination of crack length, stress intensity factors, crack tip opening displacement (CTOD), crack opening stress or load (P), etc. by means of costly apparatus and complicated computations. The measurements of crack mouth opening displacement (CMOD) and crack opening load are most important for fracture toughness (K_{IC}) calculation. Due to this evolution research for the same is carried out for precise determination fracture mechanics parameters using direct or indirect means [4]. Today's available techniques for fracture mechanics parameter calculation are costly and too demanding. To avoid all these difficulties a new technique is developed which called digital image correlation (DIC). DIC uses simple experimental set up and suitable algorithm to find out the parameter using surface strain and displacement. Digital Image Correlation (DIC) is a non-contact, non-interferometric optical technique which gives full-field displacements and strains measurements in a deformed body. The digital image correlation (DIC) technique works by comparing and matching the images taken before loading and after loading of a specimen. The use of non-contact DIC technique for calculating surface displacements and strains during mechanical testing of metallic and non-metallic materials have been investigated and studied by several researchers [5, 6,]. The detail description of this technique is found in literature [7]. The ultimate aim of the present study is to determine fracture toughness (K_{IC}) of microalloyed steel (38MnVS6) using DIC and compared their results with conventional technique.

2. Material selection

2.1 Specimen Preparation

The experiments were carried out on microalloyed steel (38MnVS6) because of its extensive use in forging industry. Yield strength and UTS of microalloyed steel are 630MPa and 935 MPa respectively. Chemical composition is as shown in table1. Fracture toughness test was carried out on compact tension (CT) specimen. The CT specimen drawing details is shown in Figure 1. The specimen surface was polished with fine abrasive paper using conventional metallographic specimen preparation method for creating suitable speckle pattern for DIC analysis. The speckle pattern can be naturally occurring or can be applied. Generally speckle pattern can be created with combination of black and white paint. First, paint the surface with a thin layer of white paint (it could be spray paint) and then apply a black mist of paint (spray paint) to create the black speckles. But in this experiment only white mist of paint was used for creating speckle pattern because specimen surface of microalloyed(38MnVS6) steel is already black so no need of black paint. The resulting speckle pattern is as shown in Figure 2.

Table1. Chemical composition of microalloyed steel (38MnVS6).

	C	Mn	Si	P	S	Cr
Comp	0.38	1.37	0.57	0.01	0.027	0.24
	Ni	Mo	Al	Cu	V	Ti
Comp	0.26	0.03	0.015	0.17	0.02	0.02

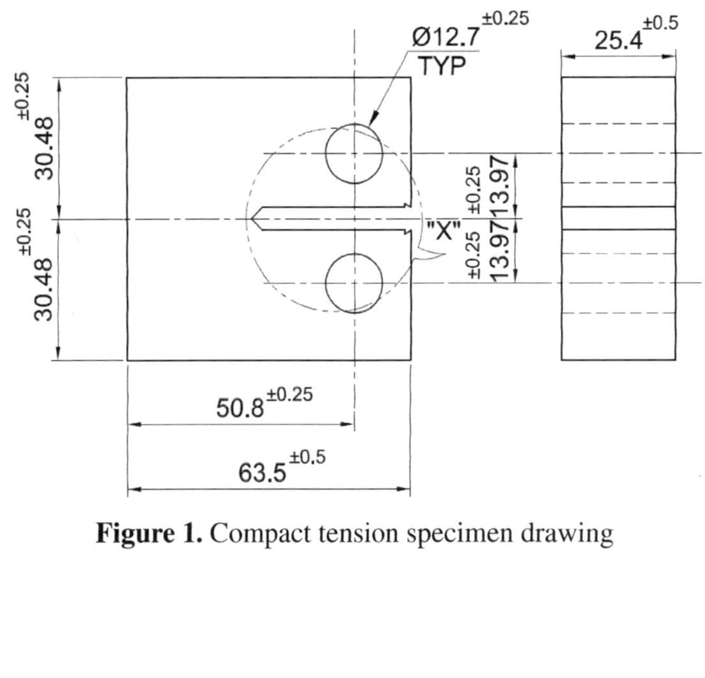

Figure 1. Compact tension specimen drawing

Figure 2. Speckle pattern created on specimen surface

2.2 Fracture Toughness Test

The fracture toughness test was carried out on Lloyed LS100K plus UTM machine. Images were taken from start to end of the experiment up to fracture. A first measured image was used as the reference image. Finally DIC was performed on each image using software 'VIC2D'. Then DIC displacement measurement gauge was placed on notch mouth to measured crack mouth opening displacement (CMOD). A Navitar 12X zoom lens was used with a Bassler CCD with maximum resolution of 1600 x 1200 pixels for image capturing. Subset size of 50 by 50 pixels was used during analysis. National Instruments DAQ System with Lab-View software was used to acquire load data during testing. Camera and lens are mounted on rigid tripod stand. Lighting was provided with halogen lamp. Figure. 3(a) shows the experimental setup of fracture toughness test using DIC. For validation of result, the test was also carried out on servo hydraulic MTS machine of capacity 250kN as shown in Figure 3(b) under same loading condition.

Figure 3. Experimental setup of fracture toughness test using,
(a) Digital image correlation (DIC) (b) Clip gauge

3. Results and discussion

3.1 Crack Mouth Opening Displacement (CMOD) Measurement

DIC extensometer is a virtual gauging tool available in software which is generally used for measurement of displacement or change in displacement between two specified points. DIC extensometer gauge (or displacement gauges) was used for CMOD measurement. DIC extensometer gauge is placed at crack mouth to measure crack mouth opening displacement (CMOD) as shown in Figure 4.

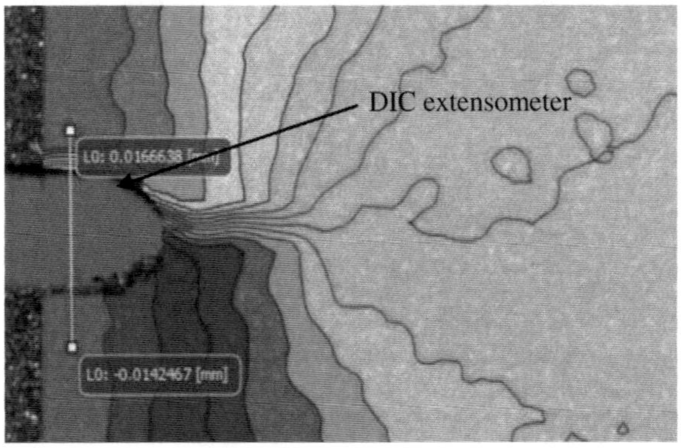

Figure 4. DIC displacement gauge for crack mouth opening displacement (CMOD) measurements

3.2 Fracture Toughness Calculation

Load (P) and crack mouth opening displacement (CMOD) data was recorded for determination of P_Q load. K_Q was calculated from P_Q by taking 5% secant and after running all validity checks given in ASTM E-399 [8], the K_Q was declared K_{IC}. It was found to be 61 MPa.m$^{1/2}$. Figure 5 shows load (P) vs. crack mouth opening displacement (CMOD) plot obtained from DIC. Fracture toughness (K_{IC}) value was also calculated using conventional method where crack opening displacement measurement was done with clip gage. The value of fracture toughness (K_{IC}) was found to be 58 MPa.m$^{1/2}$ using clip gage measurement. The observed result was quite logical therefore it is suggested that the DIC technique is feasible for fracture toughness testing.

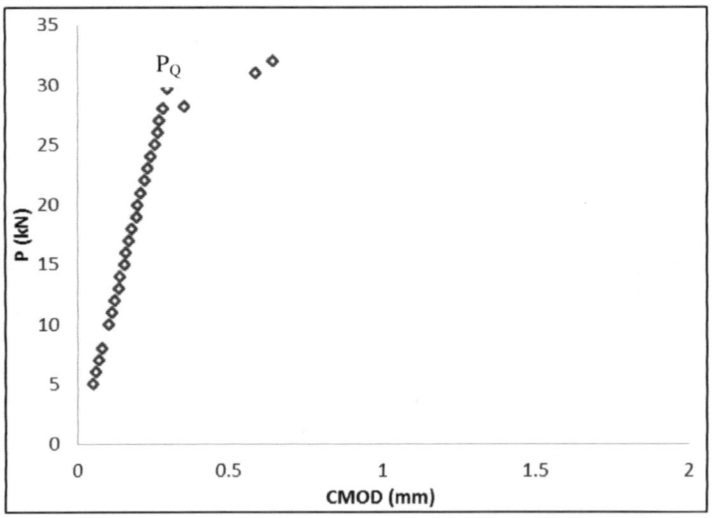

Figure 5. Load (P) vs. crack mouth opening displacement (CMOD) plot obtained from DIC

4. Conclusion

In current research the capability of digital image correlation (DIC) technique to determine the fracture toughness (K_{IC}) of microalloyed steel (38MnVS6) has been studied. Existing work proved the practical application DIC for calculating fracture toughness (K_{IC}) of microalloyed steel (38MnVS6). Fracture toughness value of micro alloyed (38MnVS6) steel via DIC method was found to be 61 MPa.m$^{1/2}$, where the same was found to be 58 MPa.m$^{1/2}$ via clip gauge technique. It can be concluded that the simple DIC technique can be a very useful and an economical substitute against clip gages.

5. Acknowledgments

The authors gratefully acknowledge the extended support provided to this work by KCTI (Kalyani Centre for Technology & Innovation) for providing financial funding, laboratory and library facilities. The authors also acknowledge the support provided by Bharat Forge Ltd, Pune. Finally, the authors would like to express special thanks and gratitude to review committee and top management of Bharat Forge Ltd for granting the permission to publish/present the research work.

6. References

[1] D.K. Matlock, G. Krauss, and J.G. Speer: Materials Science Forum Vols. 500-501 (2005), p. 87-96.
[2] A. Ktari, N. Haddar and H. Ayedi: Engineering Failure Analysis Vol.18 (2011), p. 1085-1093.
[3] X. Zhu, J. A. Joyce: Engineering Fracture Mechanics Vol. 85 (2012), p. 1–46.
[4] I. Hwang: Measurement Sci. Tech. Vol. 3 (1992) p.62-74.
[5] M.A. Sutton, W.J. Wolters, W.H. Peters, W.F. ranson, S.R. McNeill: Image vision computing Vol.1 (1983), p.133- 139.
[6] M. Zeeshan, F. Tariq N. Naz: Astronautical Congress, Cape Town, SA. (2011).
[7] J. Abanto-Bueno and J. Lambros: Engineering Fracture Mechanics, Vol. 69(2002), p.1695-1711.
[8] ASTM E 399-09 ASTM International.

Multipass-friction Stir Processing (MFSP) of Ti-6Al-4V Alloy and Investigation of Flow Properties

Sandip Chougule[1], Digvijay Sheed[1], N Prabhu[2], B P Kashyap[2], Kaushal Jha[3] and R K P Singh[1]

[1] Bharat Forge, Kalyani Centre for technology and innovation, Pune, 411036, India
[2] Department of Metallurgical Engineering and Materials Science,Indian Institute of Technology Bombay, Mumbai, 400076, India
[3] Engineering design and development division, BARC, Mumbai, 400085, India

E-mail: sandip.chougule@bharatforge.com

Abstract: Multipass friction stir processing (MFSP) of the Ti-6Al-4V alloy was carried out at 600 tool rpm and 80 mm/min traverse speed. After first pass, the initial elongated α structure transformed to prior β grains, consisting of a mixture of acicular α'and very fine lamellar α colonies along with α layer grain boundary in stir zone (SZ). This subsequently transformed to equiaxed α grain via dynamic recrystallization (DRX) process. With the increase in the number of FSP passes the fraction of equiaxed α grains was found to increase, reaching almost fully equiaxed α structure in SZ upon completion of the fifth pass. Flow properties of MFSP Ti-6Al-4V alloy were investigated by differential strain rate test carried out at 927°C. There appears no significant variation in the strain rate sensitivity index ($m \geq 0.3$) values between as received Ti-6Al-4V alloy and MFSP Ti-6Al-4V alloy specimens.

1. Introduction

Friction stir processing (FSP) is a solid state processing technique based on the basic principle of friction stir welding (FSW). In recent years, FSP technique is becoming popular along with FSW for many applications, which include producing surface composites, microstructural refinement and homogenization, microstructural modification of metal matrix composites, superplasticity and mechanical property enhancement [1-12]. Up till now, low melting point materials like aluminum alloys, copper alloys and magnesium alloys have been successfully processed via FSW/FSP process [7-9, 13-16], but it faces difficulties with the high melting point materials like titanium alloys, inconel alloys and steel [1]. To overcome these difficulties researchers successfully tried out with different advanced tool materials like tungsten-lanthanum, tungsten-rhenium, tungsten carbides, polycrystalline cubic boron nitride (PCBN) etc. [1,17].

Titanium alloys, especially, the Ti-6Al-4V alloy has been very attractive material for aerospace, bio-medical, marine and petrochemical industries because of its high strength to weight ratio and good corrosion resistance properties [18,19]. Generally, this alloy is joined by using the conventional welding processes like gas tungsten arc welding (GTAW), plasma arc welding, electron beam welding (EBW) etc. However, these processes cause the problems of melting and solidification, large thermal distortion and residual stresses, generated during fusion welding. In order to overcome these problems the friction stir welding technique is found to be a successful one [20]. The success of this technique was reported in Ti-6Al-4V alloy plates ranging from thickness of 1.5 mm to 15 mm of different starting microstructures [20-31]. The various combinations of tool rotating speed and tool traverse speed were employed in these studies. It was shown that the weld quality or processed surface was mostly affected by tool rotating speed and tool traverse speed. Also, there observed during FSW/FSP are the different process defects like tunnelling voids, flash formation and porosity [2, 14, 16, and 32]. Hence one needs to do experiment with these processing parameters prior to employing the technique for final product. In earlier studies [10-12, 33], significant grain refinement was reported in SZ in the materials undergone FSP. The fine grain size is the most important factor for superplastic behaviour of material which can be achieved by different conventional thermo mechanical processing but it comes with some limitations. In view of the several evidences that FSP can result in grain refinement of bulk materials with an enhanced superplastic behaviour, the Ti-6Al-4V alloy needs to be investigated by friction stir processing and subsequent characterizing for its superplasticity [34, 35].

In present work, MFSP of the Ti-6Al-4V alloy was carried out to study the effect of number of passes on microstructure evolution at each step. Flow properties of as received Ti-6Al-4V alloy and the same upon being subjected to MFSP were evaluated from tensile tests conducted at 927°C by differential strain rate test technique over strain rates 1 x 10-4 s-1 to 5 x 10-2 s-1.

2. Material and Experimental work

The material used in this study is mill annealed Ti–6Al–4V alloy in the form of plate of thickness 6 mm. The chemical composition (weight %) is Al 6.29, V 3.96, C 0.007, Fe 0.18, N 0.001, O 0.18 and balance Ti. Considering the low thermal conductivity and high strength of Ti-6Al-4V alloy, the tungsten based tool alloyed with 1 wt. % lanthanum oxide (W-1%L2O3) was used for FSP. The tool design had a shoulder of 25 mm diameter and a tapered pin of length 5.5 mm with the major diameter being 6 mm. In the present work, the tool head was stationary and was tilted at 2° in the traveling direction. Tilt angle of tool was adjusted by indexing the tool head. In MFSP, six passes of FSP were carried out in ambient atmosphere.

After each pass, the specimens were cut along transverse direction of the processed plates for microstructure characterization. Metallographic specimens were prepared on automatic polishing machine Tegramin-30. Etching was done by Kroll's reagent. Microstructure was examined by optical microscopy (OM) and Scanning electron microscopy (SEM). For orientation image microscopy (OIM), the specimens were electro polished in 600 ml methanol, 360 ml ethylene glycol and 60 ml perchloric acid (HClO4) solution for 20 seconds at 18 V.

Hardness was measured on the Vickers hardness tester using diamond indenter. To investigate the flow properties by using differential strain rate tests technique [35], sub-size test specimens were machined from MFSP Ti-6Al-4V alloy plates. Differential strain rate tests were performed by using 100 kN capacity, computer controlled servo hydraulic Zwick-Roell Amsler Universal Testing Machine. Tensile specimens were heated to test temperature of 927°C and soaked for 5 minutes in a three in a three zone split furnace before deformation. Then, the test specimens were deformed to total strain of 50% at nine different strain rates varying from 1 x 10-4 s-1 to 5 x 10-2 s-1. Such test was continued on the same samples with the successive strain rate cycling till the failure of test specimens. After deformation, the specimens were water quenched.

3. Results and discussion

3.1 As-received Ti-6Al-4V microstructure

The as-received Ti-6Al-4V alloy (Mill annealed) used in this study contained elongated α structure with β phase at the grain boundary as shown in Figure 1.

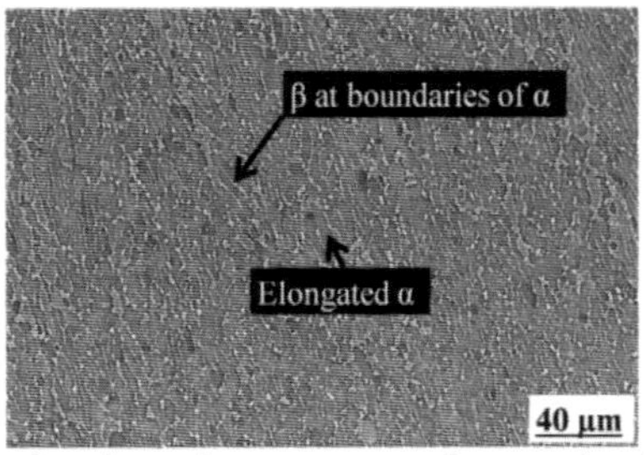

Figure 1. SEM image showing microstructure of as-received Ti-6Al-4V alloy

Total fraction of alpha phase (α) is 0.927 while that of beta phase (β) is 0.072.The average grain diameter is 4 μm observed for as-received Ti-6Al-4V alloy.

3.2 Microstructure evolution:

OIM micrograph in Figure 2 clearly reveals that as received Ti-6Al-4V microstructure consists of α phase and β phase with random grain orientation. Very less amount of recrystallized β grain fraction is observed in first pass stir zone (SZ) but, upon increasing the number of passes, the fraction of recrystallization is also noted to increase. Finally upon fifth pass, almost whole SZ showed the recrystallized α grains. After MFSP also, random grain orientation is observed in all the passes. With the increase in strain with increasing FSP pass most of α grain boundaries become high angle boundaries, which can be clearly observed by red color identification in grain boundary mapping image in Figure 3. This clearly reveals that dynamic recrystallization occurred in β phase region of the multipass friction stir processed specimen. In previous studies, it was found that, in Ti-6Al-4V alloy during deformation in α-β regime, some low angle α grain boundaries in α plate existed. This indicates the occurrence of dynamic recovery during deformation. Some dynamically recrystallized α grains of high angle grain boundaries were also reported by Furuhara et al. [36].

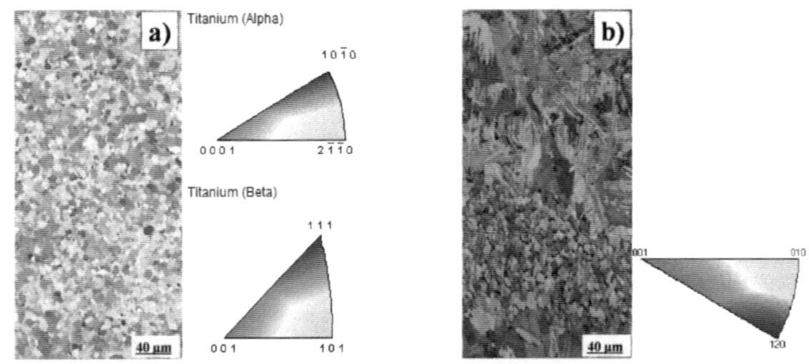

Figure 2. Multipass friction stir processed micrographs along with the inverse pole figure (a) As received Ti-6Al-4V alloy (b) First pass, (c) Third pass and (d) Fifth pass.

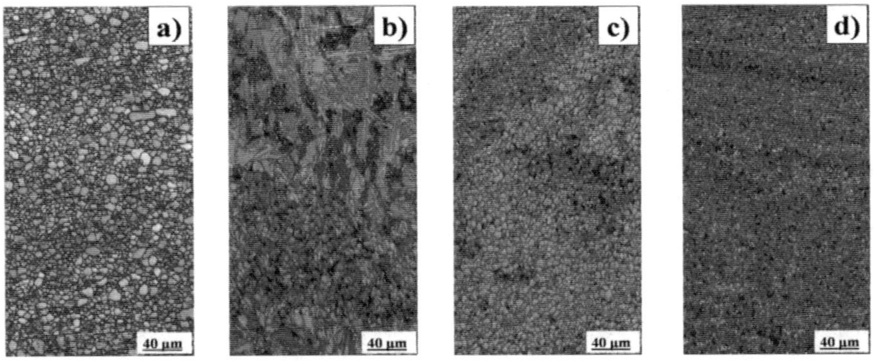

Figure 3. High angle grain boundary map of (a) As received Ti-6Al-4V alloy (b) First pass, (c) Third pass and (d) Fifth pass. (Blue color represents low angle grain boundaries whereas red color represents high angle grain boundaries)

Figure 4 shows the variation in average grain size as a function of number of FSP passes, starting from first to fifth pass. It is well understood from the plot that the grain size of material goes on decreasing with increasing number of passes. For first pass, the average grain diameter is 2.9 µm and it goes down to 1.4 µm for fifth pass. This may be due to dynamic recrystallization (DRX) that was observed from second pass till the fifth pass. The plot also shows spread of grain size distribution by the error bar line given.

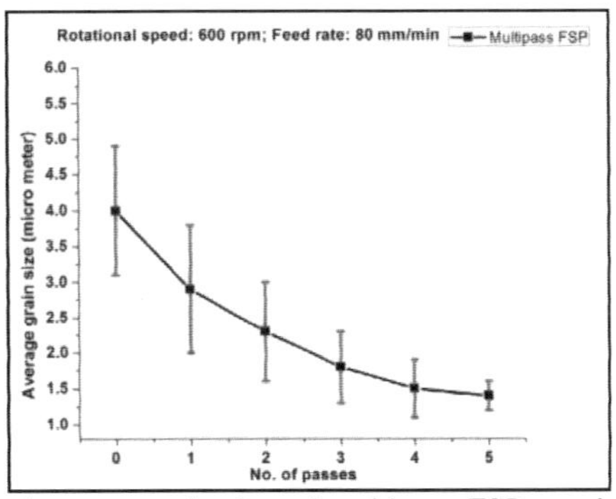

Figure 4. Average grain size of multipass FSP specimens.

3.3 Variation in flow stress during differential strain rate test:

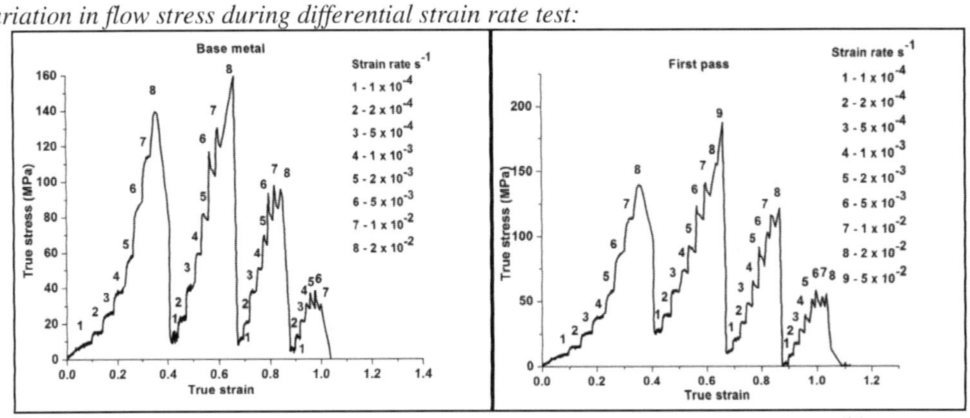

(a) (b)

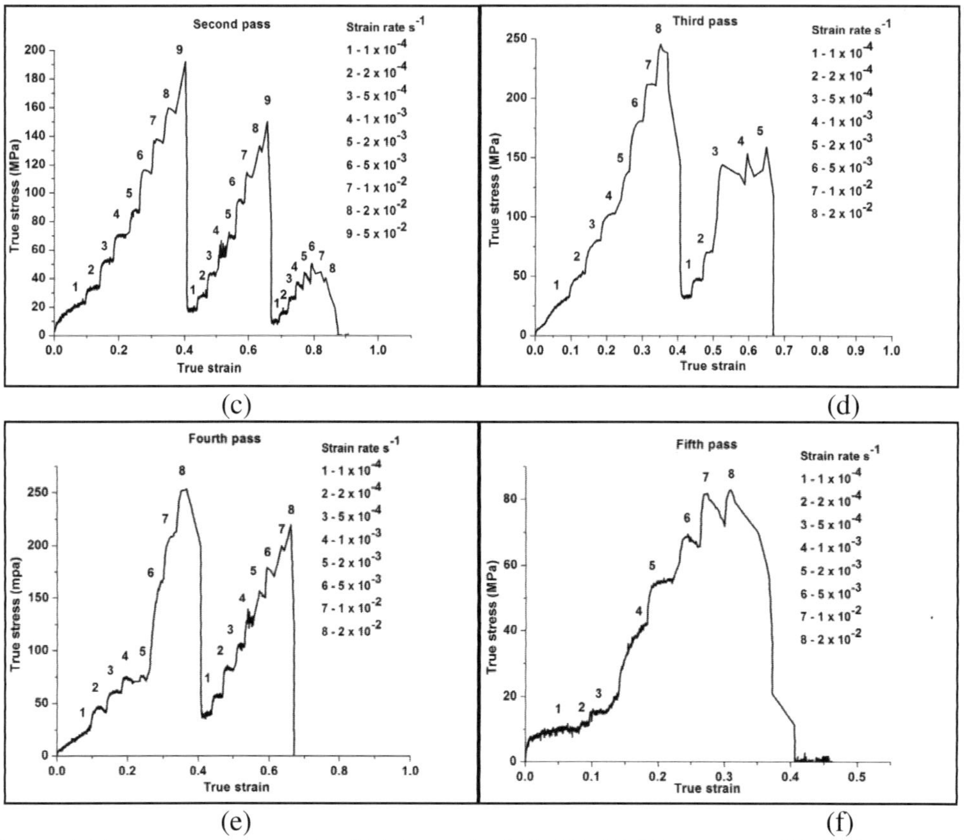

Figure 5. True stress vs true strain of (a) As received Ti-6Al-4V (b) First pass FSP (c) Second pass FSP (d) Third pass FSP (e) Fourth pass FSP (f) Fifth pass FSP.

Using the load vs elongation data from differential strain rate test, true stress vs true strain plots are given in the Figure 5 (a-f), for base metal and MFSP specimens. The tests were done from the lowest strain rate of 1 x 10-4 s-1 to the highest strain rate of 5 x 10-2 s-1. On completion of this increasing sequence of strain rate changes, the test was continued but again from the lowest to the highest strain rates. Each series of this sequential increment in strain rate test is termed as a cycle. This way, the samples were deformed as long as they could be without immature necking. The plots are given for a number of strain rate cycling up to the failure of samples. As the number of FSP passes increased, the number of differential strain rate test cycles sustained by specimen was found to reduce. The reason for this could be that as the number of passes increases there occurs an increase in porosity as well as increase in the crack length, which could lead to failure at lower true stress and smaller strain values. This is evident from the differential strain rate test conducted in the sample taken from the fifth pass.

Based on true stress vs true strain plots, ln (flow stress) vs ln (strain rate) were plotted for base metal and MFSP specimens as shown in Figure 6. Superplasticity index, identified by strain rate sensitivity index- $m \geq 0.3$, was calculated from the ln (flow stress) vs ln (strain rate) plot for base metal and MFSP specimens. For as received Ti-6Al-4V alloy and first pass friction stir processed specimens, 'm' values are found to be almost same ($m = 0.5$). As the number of passes increases the flow stress required for deformation increased significantly up to third pass. A decrease in the flow stress is observed for fourth and fifth passes. The 'm' value for second and third passes reduced up to 0.43 at lower strain rates and up to 0.2 at higher strain rates. For the fifth pass, the higher m value of 0.66 is obtained at lower strain rate. Overall, there is successive increase and decrease in the flow stress for MFSP specimens. This successively changing behavior of Ti-6Al-4V alloy MFSP specimens may be due to grain growth or dislocation strengthening of material in second and third passes. The continued deformation beyond this exhibited flow softening, which could be attributed to the mechanisms like recrystallization or recovery process, that might have occurred in the fourth and fifth pass specimens. Also, the 'm' values were also found to successively decrease and increase during MFSP. This may be because of the differences in α grain morphology developed during MFSP. Microstructure of as received Ti-6Al-4V alloy shows partial elongated grains and it contains 7 % of β phase fraction. The β phase fraction has significant effect on the strain rate sensitivity 'm' in Ti-6A1-4V alloy [37].

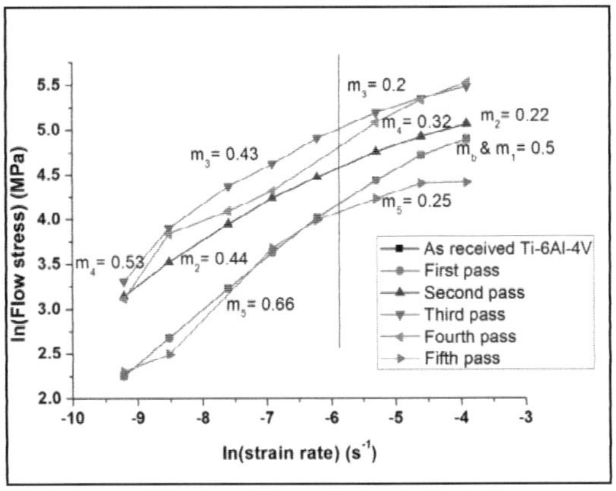

Figure 6. ln(flow stress) vs ln(strain rate) of base metal, first pass FSP, second pass FSP, third pass FSP, fourth pass FSP, fifth pass FSP

The grain coarsening of β phase is prevented through partitioning by α phase. This causes less grain coarsening and so the microstructure remains stable during deformation. In as received Ti-6Al-4V and MFSP samples, at higher strain rate, deformation is almost controlled by dislocation climb mechanism (n = 1 /m = 4), since n is stress exponent and is represented as inverse of m (n = 1 /m). On the other hand, at lower strain rate, deformation is almost controlled by grain boundary sliding (GBS) mechanism as elaborated in superplastic literature [33]. The fine grain size (<10μm) is the most important factor for superplastic behavior of material. Superplasticity is not directly dependent on strain rate, however when the strain rate is very low (approximately 2 x 10-5 s-1), prolonged exposure to high temperature causes grain growth and early failure [38].

4. Conclusions

- Microstructure evolution in the first pass revealed that the initial elongated α structure transformed to prior β grains, consisting of a mixture of acicular α' and very fine lamellar α/β colonies in stir zone (SZ). This microstructure subsequently transformed to equiaxed α with increase in the number of passes. In the fifth pass, fully equiaxed α structure was observed in SZ. In TMAZ, the prior β grains were decorated with α layer grain boundary, consisting of lamellar α/β colonies in all the passes. Heat affected zone (HAZ) was characterized by mixture of transformed β grain structure and undeformed α grains. The mechanism of microstructure evolution in SZ during MFSP is identified to be the occurrence of dynamic recrystallization (DRX) in β phase region.
- For as received Ti-6Al-4V alloy and MFSP specimens, strain rate sensitivity (m) value was found to be greater than 0.3. There occurred successive increase and decrease in flow stress for MFSP specimens. The strengthening arises from martensitic phase transformation and grain coarsening in second and third FSP passes, whereas, the subsequent softening in the fourth and fifth FSP passes occurs by dynamic recrystallization or recovery process.

5. References

[1] Mishra Rajiv S. and Z. Y. Ma. 2005 *Materials Science and Engineering: R* 50 1-78.
[2] Pilchak A. L., M. C. Juhas and J. C. Williams. 2007 *Metallurgical and Materials Transactions A* 38 401-408.

[3] Pilchak A. L., D.M. Norfleet, M.C. Juhas and J.C. Williams. 2008 *Metallurgical and Materials Transactions A* 39 1519-1524.
[4] Sharma S. R., Z. Y. Ma and Rajiv S. Mishra.2004 *Scripta Materialia* 51 237-241.
[5] Kwon Y. J., N. Saito and I. Shigematsu. 2002 *Journal of Materials Science Letters* 21 1473-1476.
[6] Ma Z. Y., S. R. Sharma and Rajiv S. Mishra. 2006 *Scripta materialia* 54 1623-1626.
[7] Johannes L. B. and R. S. Mishra. 2007 *Materials Science and Engineering: A* 464 255-260.
[8] Rao A. G., V.A. Katkar, G. Gunasekaran, V.P. Deshmukh, N. Prabhu and B.P. Kashyap 2014 *Corrosion Science* 83198-208.
[9] Feng A. H. and Z. Y. Ma2007 *Scripta materialia* 56 397-400.
[10] Charit I. and Rajiv S. Mishra. 2003 *Materials Science and Engineering: A* 359 290-296.
[11] Ma Z. Y., R.S. Mishra, M.W. Mahoney and R. Grimes. 2003 *Materials Science and Engineering: A* 351 148-153.
[12] Ma Z. Y., Rajiv S. Mishra and Murray W. Mahoney. 2002 *Acta materialia* 50 4419-4430.
[13] Liu G., L.E. Murr, C-S. Niou, J.C. McClure and F.R. Vega. 1997 *Scripta materialia* 37 355-361.
[14] Park Hwa Soon, Takahiro Kimura, TaichiMurakamic, Yoshitaka Naganod, Kazuhiro Nakata and Masao Ushio. 2004 *Materials Science and Engineering: A* 371 160-169.
[15] Su Jian-Qing, T.W. Nelson, T.R. McNelley and R.S. Mishra. 2011 *Materials Science and Engineering: A* 528 5458-5464.
[16] Lee Won-Bae, Jong-Woong Kim, Yun-Mo Yeon and Seung-Boo Jung. 2003 *Materials Transactions* 44 917-923.
[17] Rai R., A. De, H. K. D. H. Bhadeshia and T. DebRoy. 2011 *Science and Technology of welding and Joining* 16 325-342.
[18] Lütjering Gerd and James Case Williams. 2003 *Titanium*. Vol. 2. Berlin: Springer.
[19] Leyens Christoph and Manfred Peters. 2003 *Titanium and titanium alloys*. Wiley-VCH, Weinheim.
[20] Su Jianqing, Rajiv S. Mishra, JiyeWang, RayXu and JohnA.Baumann. 2013 *Materials Science and Engineering: A* 573 67-74.
[21] Davies P. S., B.P. Wynne, W.M. Rainforth, M.J. Thomas and P.L. Threadgill. 2011 *Metallurgical and Materials Transactions A* 42 2278-2289.
[22] Edwards P. D. and M. Ramulu. 2009 *Science and Technology of Welding and Joining* 14 476-483.
[23] Leo Paola and Emanuela Cerri. 2014 *Materials Science Forum*. Vol. 783 574-579
[24] Zhang Yu, Yutaka S. Sato, Hiroyuki Kokawa, Seung Hwan C. Park and Satoshi Hirano. 2008 *Materials Science and Engineering: A* 485 448-455.
[25] Liu H. J., L. Zhou and Q. W. Liu. 2010 *Materials & Design* 31 1650-1655.
[26] Zhou L., H. J. Liu and Q. W. Liu. 2010 *Materials & Design* 31 2631-2636.
[27] Pilchak A. L., W. Tang, H. Sahiner, A.P. Reynolds and J.C. Williams. 2011 *Metallurgical and Materials Transactions A* 42 745-762.
[28] Ramirez Antonio J. and Mary C. Juhas. 2003 *Materials Science Forum*. 426 2999-3004.
[29] Edwards P. and M. Ramulu. 2010 *Journal of Engineering Materials and Technology* 132 031006.
[30] Edwards P. and M. Ramulu. 2010 *Science and Technology of Welding & Joining* 15 468-472.
[31] Ma Z. Y. 2008 *Metallurgical and Materials Transactions A* 39 642-658.
[32] Chen Hua-Bin, Keng Yan, Tao Lin, Shan-Ben Chen, Cheng-Yu Jiang and Yong Zhao. 2006 *Materials Science and Engineering: A* 433 64-69.
[33] Nieh Tai-Gang, Jeffrey Wadsworth and O. D. Sherby. 2005 Cambridge university press.
[34] Kashyap B. P., A. Arieli and A. K. Mukherjee. 1985 *Journal of materials science* 20 2661-2686.
[35] Kashyap B. P. and G. S. Murty. 1983 *Journal of Materials Science* 18 2063-2070
[36] Furuhara Tadashi, B. Poorganji, H. Abe and T. Maki. 2007 *Jom* 59 64-67.
[37] Sieniawski J. and M. Motyka. 2007 *Journal of Achievements in Materials and Manufacturing Engineering* 24 123-130.

[38] Arieli A. and Rosen A. 1977 *Metallurgical transaction A* 8 1591-1596

IOP Conf. Series: Materials Science and Engineering **422** (2018) 012018 doi:10.1088/1757-899X/422/1/012018

Analysis on Stress Field of Asymmetrical Cold Ring Rolling for High-speed Rail Bearing Inner Ring

J Y Xu[1,2] and B S Sun[1,2]

[1] Faculty of Mechanical Engineering and Mechanics, Ningbo University, Ningbo 315211, People's Republic of China

[2] Zhejiang Provincial Key Lab of Part Rolling Technology, Ningbo University, Ningbo 315211, People's Republic of China

Email: sunbaoshou@nbu.edu.cn

Abstract. In this paper, the finite element model of asymmetric cold ring rolling for high-speed rail bearing inner ring is built by using ABAQUS software and the inner stress field of the ring metal during the whole cold ring rolling process is researched and analyzed, which reveals. The distribution of radial , axial , circumferential and equivalent stress at the bite rolling stage, the stable rolling stage and the fine rolling stage in the asymmetric cold rolling forming process for high-speed rail bearing inner ring were analyzed respectively. The stress in each direction of the ring was found to increase first and then decrease with each stage of cold rolling going. Finally, it tends to be stable. The stress on the contact area between the ring blank and the drive roll and the mandrel is obviously greater than that on the core of ring, and the stress in the core is more uniform.

1. Introduction

Ring cold rolling forming technology is an advanced manufacturing technology for the production of various seamless ring parts such as bearing rings, high-neck flanges, railway wheels and marine turbine rings [1,2]. It has been widely used in manufacturing fields such as automobiles, trains, ships, machinery manufacturing, aerospace, petrochemicals, and atomic energy because of its advantages of saving material, high processing precision, high reliability and high production efficiency [3]. Many scholars have done a lot of researches in this field. Hua et al. [4] conducted the finite element simulation on the cold ring rolling of the ring section of the groove ball by using ABAQUS software, studied the variation of roundness and optimized the size and process parameters of the ring blank. Yang et al. [5] studied the relationship between the ring radius expansion deformation and the press quantity of a typical profiled ring member which is a stepped ring ring during the cold roll forming process. Mao et al. [6] simulated and optimized the feeding speeds of a set of blanks and core rolls, analyzed and compared the effects of different width-radius ratios and feed speeds on the rolling force and rolling torque, and obtained the optimal parameters. Jin et al. [7] used the test to verify the reliability of the simulated cold ring rolling process. The simulation results were in good agreement with the test results, which verifys the rationality of program on the optimized blank cold ring rolling. Xu et al. [8] proposed a new mathematical model for predicting the diameter expansion of the flat ring in the radial–axial rolling proces. Li et al. [9-10] investigated deeply the effect of the mandrel and drive roll radius on uniformity of strain in cold-closed T-shaped ring rolling process.

2. Principle of cold ring rolling

Cold ring rolling technology is a kind of cold forming process that makes the inner metal material of the ring parts undergo plastic deformation with the help of the extrusion effect of drive roll and mandrel of cold rolling mill at room temperature. The drive roll is driven by the motor to rotate at a constant speed along its central axis, and the mandrel is moved along the radial direction of the ring in the radial direction by the pressing device of the cold ring mill. Under the combined action of friction and extrusion forces, the ring rotates and plastically deforms. At the same time, its diameter gradually increases and the wall thickness gradually decreases. During this period, the guide roll always adheres to the ring to ensure its roundness. When the diameter of the ring part gradually increases to reach the measuring roll at the predetermined position, the mandrel stops the radial linear feed motion, and finally the ring part falls from the cold ring rolling mill, and the entire cold ring rolling process ends. Figure 1 shows the basic working principle of ring during cold ring rolling.

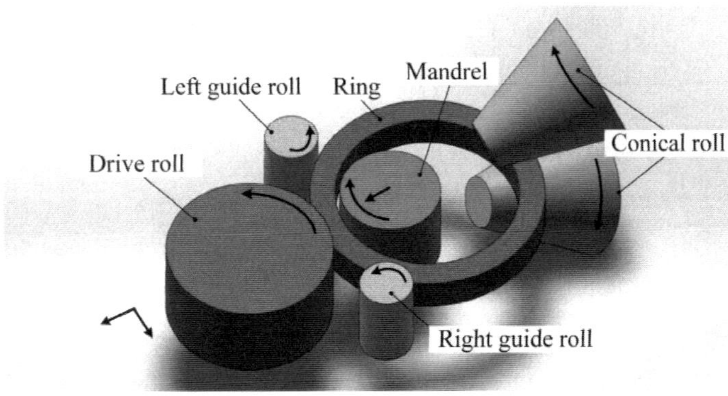

Figure 1. Basic working principle of ring rolling

3. Finite element simulation of cold ring rolling

3.1. Material

High-speed rail bearings are subjected to very large radial load and road impact load, so sevice life and performance are required to be higher. GCr15 bearing steel is selected as material of high-speed rail bearing inner ring in this paper. The comprehensive property of GCr15 bearing steel is good, which has high hardness, good wear resistance, the uniform texture and high contact fatigue strength. The chemical composition and physical properties of GCr15 bearing steel are shown in Table 1 and table 2 respectively.

Table 1. Chemical composition of GCr15 bearing steel

Element	C	Cr	Mn	P	Si	Ni	Mo	Cu	S
Percentage (%)	0.95~1.05	1.30~1.65	0.20~0.40	≤0.027	0.15~0.35	≤0.30	≤0.10	≤0.25	≤0.02

Table 2. Physical properties of GCr15 bearing steel

Material	Density (g/cm^3)	Thermal expansion(10^{-6}/)	Elastic modulus(GPa)	poisson ratio	Hardness (HB)
GCr15	7.8	11	208	0.3	800

3.2. Finite element model

The finite element model of cold ring rolling for high-speed rail bearing inner ring is built which is based on ABAQUS software, as shown in Figure 2. The main process parameters of the finite element simulation are shown in Table 3.

Table 3. Process parameters for cold ring rolling of finite element simulation

Process parameter	Value
Driving roll speed（rad/s）	16
Mandrel first feed speed（mm/s）	1
Mandrel sixth feed speed（mm/s）	0.2
Conical roll speed（rad/s）	29
Friction factor between rollers and ring	0.5
Network elements of ring	12096

Figure 2. Finite element model of asymmetric cold rolling of high speed railway bearing inner ring

4. Finite element simulation results

4.1. Stress field of bite rolling stage

At the bite rolling stage of asymmetrical cold rolling for the high-speed rail bearing inner ring, the cold ring rolling has just begun. The inner wall surface of the ring blank and the outer surface of the mandrel are in close contact with each other, and the mandrel performs linear feed movement. The drive roll rotates around its center axis. Under the effect of friction force, the ring blank rotates. At the same time, the internal metal material undergoes plastic deformation and metal flow. The diameter of the ring gradually increases, and the rolling force increases slowly and is very unstable. The simulated results of the stress in radial, axial, and circumferential direction is presented as Figure 3.

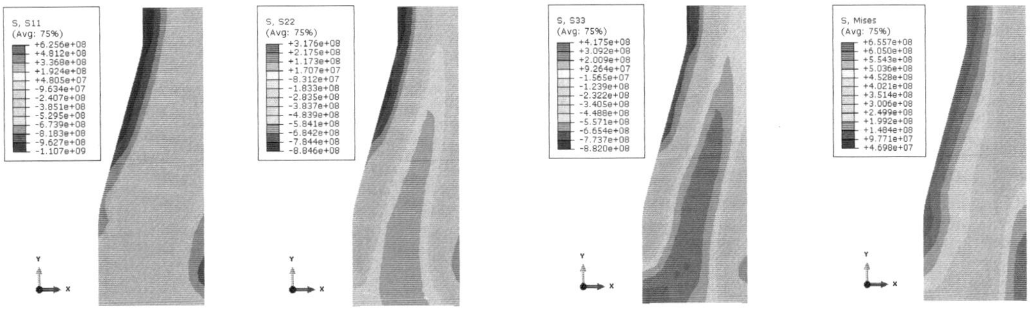

(a) (b) (c) (d)

Figure 3. Stress field of biting compression stage(a)Radial stress, (b)Axial stress, (c) Circumferential stress, (d) Equivalent stress

It can be seen from Figure 3(d) that it is mainly manifested as the contact stress between the blank and the drive roll and the mandrel at the bite rolling stage. At the same time, it can be observed that the contact stress is mainly concentrated on the inclined step and the mandrel side of lardge end face of the blank , which is directly related to the shape of the blank. Since the blank is subjected to the pressure on the side of the inclined step of the drive roll, the opposite force is generated on the side of the mandrel. Figure 3(a) shows at the same place, there is a larger contact compressive stress , which indicates that the ring is radially compressed during the cold ring rolling. Figure 3(b) shows that the surface contact area exhibits axial compressive stress and is mainly concentrated on the side of the small end face, while the blank center is mainly axial tensile stress, which explains the axial elongation at the beginning of the rolling process. In addition, a slight "fish-tail" phenomenon can be observed on the side of large end face. Expect the contact compressive stress on the surface, the center of the ring is tensile stress as shown in Figure 3(c), which indicates that there is a tendency of circumferential expansion during bite rolling stage.

4.2. Stress field at stable rolling stage

At the stable rolling stage of the high -speed rail bearing inner ring, with the rotation of the drive roll and the continuous feeding of the mandrel, the plastic deformation and the circumferential flow of the metal material inside the blank are accelerated. This will result in increase of the diameter of the ring blank but the decrease of wall thickness of the ring blank. Since the ring blank has undergone stable elasto-plastic deformation, the rolling force gradually increases and reaches the maximum value, which is the main stage of the cold ring rolling for the high-speed rail bearing inner ring. Figure 4 shows the simulated results of the stress at the stable rolling stage in radial, axial, and circumferential direction.

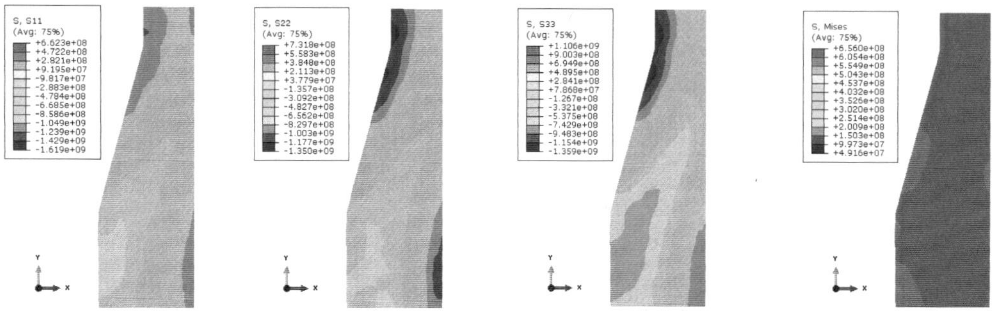

(a) (b) (c) (d)

Figure 4. Stress field at stable rolling stage(a)Radial stress,(b)Axial stress,(c) Circumferential stress,(d) Equivalent stress

From Figure 4(d), it can be observed that when it enters the stable rolling stage, the plastic deformation is the most severe, and the stress distribution of the entire section is uniform at the same time, indicating that the forming process in all areas is stable and the stress is uniform. Figure 4(a) shows that the compressive stress is dominant in the radial direction, and the concentration area of the compressive stress is basically the same as that of the bite rolling stage, and the compressive stress on the side of the large end surface of the drive roll is smaller. Figure 4(b) shows that the compressive stress is on the side of the small end face, the stress distribution of the contact surface is larger than that of the inner. The side of the large end face of the mandrel is mainly the compressive stress. This is due to side baffle of the mandrel limit axial broadening, which leads to the obstruction of the lateral metal flow and the increase of contact stress with the mandrel. It can be seen from Figure 4(c) that on the side of the drive roll in the circumferential direction, compressive stress is at the small end face and tensile stress is at the large end face, while it is opposite on the side of the mandrel. Observing the shape of the blank, it was found that the fish-tail that appeared on the large end face at the bite rolling stage gradually disappeared with the limitation of the baffle as the rolling progresses.

4.3. Stress field at fine rolling stage

At the fine stage of the asymmetrical cold ring rolling for the high-speed rail bearing inner ring, the rolling process comes to an end. With the stop of the linear feed movement of the mandrel, the mandrel is maintained at the position where the feed ends, and the ring continues to be pressed. The blank and the drive roll continue to rotate, and the blank and the mandrel are driven to rotate under the action of the friction force, and the guide roll also reaches the predetermined position and no longer follows. The effect of fine and returning round is achieved, and the forming precision of the cold ring rolling is improved. Figure 5 shows the simulated results of the stress at the fine rolling stage in radial, axial, and circumferential direction.

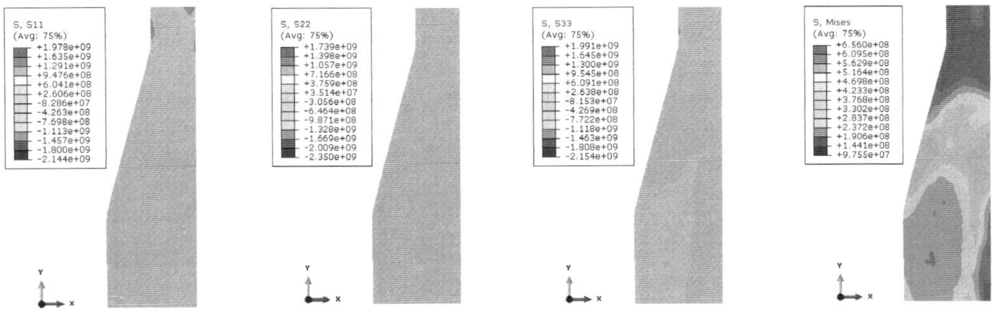

(a) (b) (c) (d)

Figure 5. Stress field at fine rolling stage(a)Radial stress,(b)Axial stress,(c) Circumferential stress, (d) Equivalent stress

It can be observed from Figure 5(d) that since the mandrel stops feeding, the stress has decreased significantly compared with the main forming stage. At the same time, the stress distribution is similar to the former stages of the rolling process, presenting two stress concentrated areas. The stress concentrated area is mainly the severely deformed area on the side of the small end face and the side of the mandrel on the large end face. Comparing the three graphs in Figure 5（a）（b）（c）with the corresponding graphs in Figure 3 and Figure 4, it is found that the stress distribution at the fine rolling stage is basically the same, the stress is concentrated on the side of small end surface, and the side of large end surface close to the mandrel. However, the difference is that the magnitude of the stress is significantly decreased, and the distribution of stress is also more uniform. This shows that the stress and the tendency of plastic deformation is smaller at the fine rolling stage.

5. Conclusion

In this paper, the distribution of radial , axial , circumferential and equivalent stress at the bite rolling stage, the stable rolling stage and the fine rolling stage in the asymmetric cold ring rolling forming process for high-speed rail bearing inner ring were analyzed respectively.

(1) At the bite rolling stage, the rolling force of cold ring rolling increases slowly and is very unstable. The stress of the ring blank mainly concentrates on the top end of the outer wall of the ring blank and the inner bottom end, and gradually expands to the metal structure of the core.

(2) At the stable rolling stage of the high -speed rail bearing inner ring, the rolling force gradually increases and reaches the maximum value. The stress distribution of the ring blank is presented as tensile stress and relatively uniform. At this stage, the stress has penetrated into the core of the ring blank. This indicates that the diameter of the ring blank increases faster, the plastic deformation becomes more severe and has permeated through the entire blank, the stress in all directions shows an increasing trend.

(3) At the fine rolling stage, the stress distribution in each direction in the inner ring blank is more uniform, the equivalent stress value tends to be stable and shows a downward trend, and the degree of plastic deformation of the metal structure gradually weakens.

6. Acknowdgements

This project was supported by Major special project of industrial technology innovation in ningbo city (2015B11045).

7. References

[1] Hua L, Huang X G and Zhu C D, 2001. *Theory and Technology of Ring Rolling* (Beijing: Machinery Industry Press)

[2] Hua L and Qian D S 2014 Ring Rolling Forming Theory and Technology for Bearing *J. Mech. Eng.* **50** 70-75

[3] Li C T and Huang X 2004 Ring Rolling Technology and Its Application in China *Forging equipment and manufacturing technology*. 8-13

[4] Hua L, Zuo Z.J, Lan J and Qian D.S 2006 Research on the Following Motion Rule of Single Guide Roller in Cold Ring Rolling *Chinese Journal of Mechanical Engineering.* **17** 1082-1086

[5] Yang H, Li L Y, Wang M and Guo L G 2010 Research on the Extended Deformation Behavior of Rings in Cold Rings with Profiled Rings *Chinese Science: Technical Science.* **40** 802-810

[6] Mao H J and Tian L 2010 Numerical simulation and experimental research on cold rolling of outer rings of double groove ball bearing *Bearing.* **2** 25- 28

[7] Jin Y Y, Qian D S and Shi L 2016 Numerical simulation and experiment on cold rolling of bearing rings with L-shaped section *Bearing*. 15-20

[8] Xu W J, Yang X B, Gong X.T and Zhou J 2012 A new mathematical model for predicting the diameter expansion of flat ring in radial–axial ring rolling. *Int J Adv Manuf Technol.* **60** 913–921

[9] Li L Y, Li X, Liu J, He Z 2016 Effects of roll sizes on uniformity of strain and variation of configuration in cold-closed t-shaped rolling process of he30 aluminum alloy ring *J Mater Process Technol.* **86** 191–201

[10] Zhu X L, Liu D, Yang Y, Hu Y H, Liu G.W and Wang Y N 2016 Effects of blank dimension on forming characteristics during conical-section ring rolling of inco718 alloy *J Mater Process Technol.* **84** 1–12

IOP Conf. Series: Materials Science and Engineering **422** (2018) 012019 doi:10.1088/1757-899X/422/1/012019

Design and Development of 3D Printed Dental Drill Guide for Better Flow of Irrigant at Drilling Site

Varun Arora[1, a], Sanjeev Kumar [2,b], Parveen Kalra[3,c] and Rahul Datta[4,d]

[1]Mechanical Engineering Department, PEC University of Technology, Chandigarh, India

[2]Mechanical Engineering Department, PEC University of Technology, Chandigarh, India

[3]Production & Industrial Engineering Department, PEC University of Technology, Chandigarh, India

[4]Datta Dental, #416, 2nd floor, Sector-37A, Chandigarh, India.

Email: [a]arora_varun84@yahoo.co.in, [b]skthakkarpec@yahoo.com, [c]kalrapec@gmail.com, [d]docdatta@gmail.com

Abstract. The aim of this study was to design and fabricate a dental drill guide with built-in external irrigation mechanism for improved flow of irrigant at the drilling site to restrict the bone temperature below the threshold of thermal necrosis i.e. 47 ^{0}C during implant site preparation. This guide was compared with the conventional type 3D printed drill guide (commercially available in the market) in terms of heat generated while drilling.

Temperature was measured with K type thermocouple and data acquisition system using PPMA as a work piece instead of bone. Constant drilling load of 1.7 Kg was applied using a dedicated experimental setup. Drill speed was kept constant at 1200 rpm throughout the experimentation. Temperature was measured at the depths of 3, 7 and 11 mm with the help of thermocouples placed perpendicular to the drill axis. Incremental drilling as prescribed the drill manufacturer was performed. More than 10% improvement in maximum temperature was recorded at all the drilling locations when new drill guide was used.

1. Introduction

In spite of noticeable developments in treatment and precaution of oral care, lots of teeth are lost. Males and females of various age groups are self conscious regarding their dentures, bridges or missing tooth. Many experience difficulty in talking as their dentures slip or click. Moreover, they feel aged before time due to tooth loss [1]. Attitude of people regarding tooth loss has begun to change [2]. Aesthetic instead of operational factors determines an individual's requirement about replacement of lost tooth. The need for the replacement is highly associated with the location of the lost tooth [3].

A variety of alternatives like bridges, dentures and dental implants are available for the replacement of missing teeth but the latest of them are dental implants. [1] Dental implants fix in a jaw bone. So, they function more like a natural tooth and becomes ideal choice for the replacement of lost teeth. Moreover, dental implants integrate with the jawbone and help in preventing bone loss. Dental implants provide an outstanding substitute to the disadvantages of traditional dentures and bridges.

The primary reason for the success of dental implant is the biocompatibility of titanium, due to which it bonds to the bone by the process of osseointegration. Biocompatibility is basically acceptance of an artificial implant by the adjacent tissues as well as by the body in total. Therefore, biocompatibility becomes an essential requirement for the implant along with other functional requirements [4]. Titanium and its alloys are well acceptable by the body and it also meets the other functional requirements.

In cases of a tooth loss in a patient, having limited bone to support the implant and due to adjacent anatomical structures drilling has to be performed very carefully, preferably with help of dental drill guides. These guides have improved patient care by allowing implant placement in areas that may have required grafting and additional surgeries. Drill guides acts a navigator and help surgeons to place implants more accurately and clearly [5].

Metallic sleeves inside the 3D printed template facilitate the surgical procedure by guiding the drill to the exact location and orientation as determined in the software. Drill guide needs to fit perfectly on the mating surface of the jaw because of the complicated shape of the bone. Surgical guides can be manufactured using 3D printing techniques like stereolithography, polyjet or FDM and are custom manufactured for each patient [6].

Bone temperature is considered the most vital parameter while performing dental surgery and it must be maintained below the starting temperature of necrosis, i.e. 47^0C. Thermal conductivity of cortical bone is 0.16–0.34w/m/k and that of cancellous bone is 0.30w/m/k, this implies that the heat produced while drilling is preserved at the cortical portion the bone. Taking into consideration the lower thermal conductivity of bone, drilling could cause a cumulative heat increase, particularly in the cortical bone. That's why, the cavity needs to be irrigated regularly throughout drilling process. Irrigation can be external or internal. External irrigation seems to be more efficacious in comparison to internal irrigation because firstly, in internal irrigation there are chances of clogging of holes provided for irrigation in the drill bit due to bone chips. Secondly, the cortical bone is tougher and has higher coefficient of friction than the cancellous bone. It experience frictional forces for a longer period of time compared to the deeper regions of the cavity [7].

Conventional dental drill guides have a drawback of rise in bone temperature. These guides must fit and adapt perfectly in order to achieve accuracy of implant placement and this can prevent adequate irrigant from reaching the osteotomy site which increases the temperature above 47^0C. It further results in problems like bone cell death, longer healing time and even failure of implants. Therefore, it becomes extremely important to design and develop dental drill guides which are not only accurate but also provides adequate irrigation at drilling site.

2. Material and Methods

2.1. The Design

Existing dental drill guides rely on manual external cooling via syringe or tubes attached to the hand piece. Therefore, in this study, an attempt has been made to design and develop a dental drill guide having improved external irrigation system as inbuilt feature to eliminate the negative aspect of existing dental drill guide.

A dental drill guide basically comprises of two components: sleeves/ bushes (which act as a guiding cylinder) and the contact surface or template (Fig. 1). Sleeves are generally made from surgical grade stainless steel whereas templates are fabricated by different 3D printing techniques.

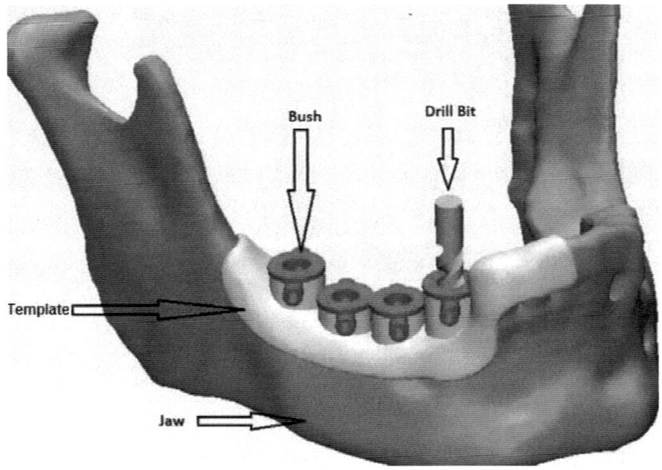

Figure 1. Schematic arrangement of 3D printed dental drill guide placed over a Mandible

The objectives of this research work are as follows:

- Better circulation of irrigant to the control bone temperature
- Precise fitment of guide over the jaw
- Accuracy of drilled holes
- Easily adaptable by the surgeons
- Easy to fabricate

2.1.1. Design and Development of Template. Template is 3D printed part which adapts the bone or the tissue of the patient depending upon the type of guide. In this case bone supporting guide have been created. Therefore, it has to adapt very precisely above the jaw bone.
Mimics Innovation Suite (MIS) was used for designing. It is an image processing software commonly used for studying human anatomy and designing custom specific biomedical equipments (like guides). Designing process involves segmentation and 3D model creation of mandibles in Mimics followed by virtual implant placement and template design in 3-matic. After designing, the template was 3D printed on Eden 260V (a polyjet technology based 3D printing machine by Stratasys).

2.1.2. Designing and Development of Bush / Sleeve. To increase the circulation of irrigant at the drilling site for controlling the bone temperature, irrigation channels were provided in the internal diameter of bush through grooves. The supply of irrigant to these irrigation channels is through inlet pipes brazed in the transverse direction of the bush (Fig. 2). In this new method of irrigation, fluid flow is from two directions which improve the fluid flow for cooling and removal of bone debris / chips where as in the conventional design these bushes do not have any such irrigation channels and fluid flow may get restricted.

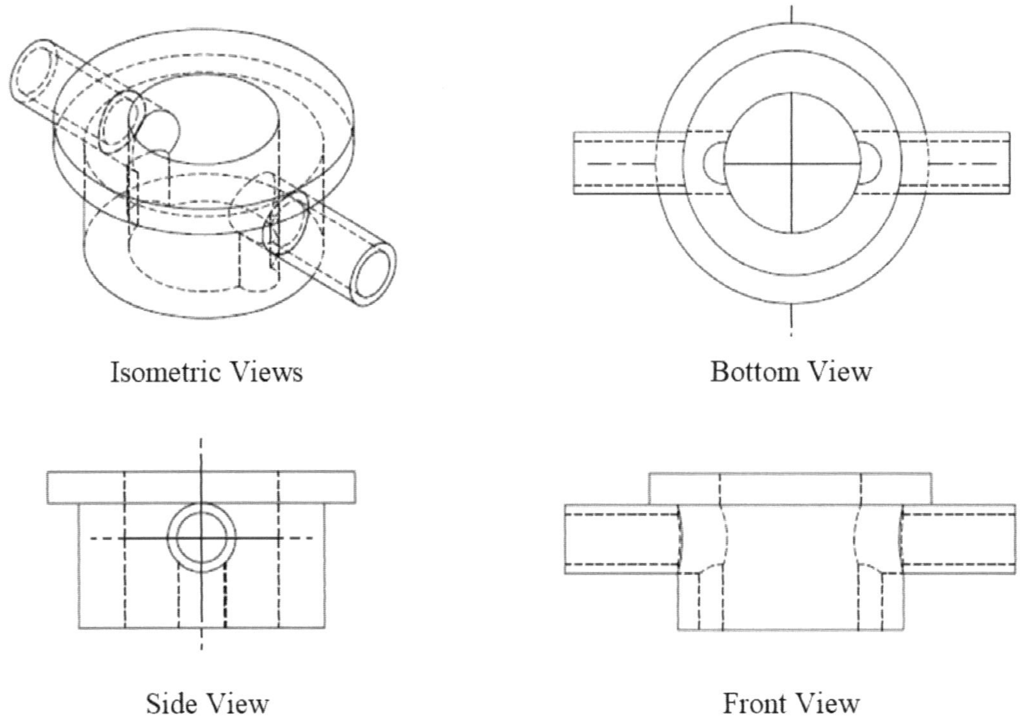

Figure 2. Different views of improved bush / sleeve, used in dental drill guide

2.2. Evaluation

In order to evaluate the new design in terms of rise in bone temperature while executing dental implant surgery, comparative in-vitro experimentation was performed. With the intention to execute the trials in an affordable manner and to nail down impact of homogeneity and anisotropy of bone on the results, polymethylmethacrylate (PMMA/Acrylic) was chosen for the experimentation over the bovine bone. PMMA is a homogeneous and isotropic material with properties near to that of bone, particularly the thermal conductivity. Thermal conductivity (W/mk), specific heat (J/kg K), density (kg/m^3) and thermal diffusivity (m^2/s) of (bovine cortical bone, femur/PMMA) are 0.54/0.50, 1260/1297.9, 1800/1683, 2.38x10^{-7}/2.29x10^{-7} respectively [8].

2.2.1. Experimental Apparatus. With a purpose to effectively evaluate the performance of newly designed guide, it was required to control drilling parameters like rpm, feed, coolant flow rate, torque and pressure / load on the hand piece etc. Almost all the listed variables can be controlled except drilling pressure because it depends upon the surgeon and it may vary as the surgical procedure continues. Therefore, a dedicated system was created to carry out the testing (Fig. 3). With the aid of this system it was possible to control the pressure.

Temperature changes during the drilling sequence were assessed with K type thermocouple having 2 mm tip diameter, 2 m long and temperature range from 0 ^{0}C to 482 ^{0}C. To record the temperature at different locations, three thermocouples were placed in the wells made in the work piece. Thermocouples were connected to the National Instrument's data acquisition system (NI 9214), it is a 16 channel thermocouple module having sampling rate of 68 samples per second. This DAQ was further connected to a computer having Lab view software. Data received from the thermocouples was digitized and saved in the computer after processing through data acquisition system and Lab view software.

Figure 3. Equipment designed for experiments

2.2.2. Sample Preparation and Experimentation. Pieces of 40 mm length were sliced from acrylic rod of 25 mm diameter. It was decided to measure the temperature at depths of 3, 7 and 11 mm from the top surface. To do this, three wells were created in acrylic work piece and thermocouples were placed in these wells. The wells were drilled perpendicular to axis of drill pathway. Total of six samples were made, three for performing the drilling using conventional drill guide and three for performing the drilling using newly designed drill guide.

Two hours prior to starting the experiments, all the instruments including work piece and saline (used as irrigant) were kept in an air conditioned room at 27^0C and the testing was performed at the same temperature. The prepared acrylic samples were mounted on a fixture of drilling unit (made to perform the drilling operation) and thermocouples were positioned in the wells filled with thermal paste.

The setup for external irrigation with conventional guide versus external irrigation with newly designed guide is shown in Fig. 4. Each set of experimentation was repeated three times. Irrigant was supplied at constant flow rate with the help of physiodispenser pump and the depth of the drilling was 13 mm. The implant cavity was prepared with contra angled hand piece at a (rotational speed 1200 rpm, contact pressure 1.7 kg and torque 20 Nm).

To simulate clinical operating situation, incremental drilling was carried out in the same cavity. During the drilling process, breaks were taken to change the drill bits and the site was continuously irrigated in that duration. The time taken to change the drill is cool-down time and it's a standard practice. Data regarding the changes in temperature was logged in a PC using NI data acquisition system and Lab View software.

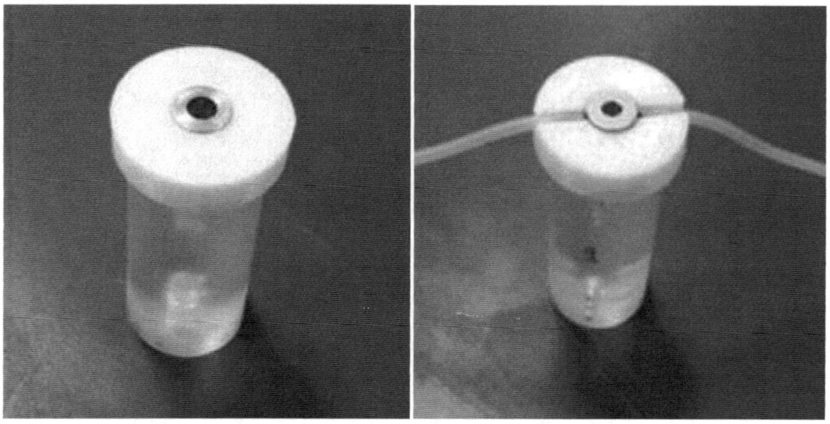

Figure 4. Setup for conventional and modified drill guides

3. Results and Discussions

The comparison of guides in terms of temperature for newly developed dental drill guide (having better circulation of irrigant at the drilling site) and conventional drill guide is given in Table 1.

Table 1. Comparison of conventional and modified drill guides

Depth	Conventional drill guide		Modified drill guide	
	Average temperature [^{O}C]	Maximum temperature [^{O}C]	Average temperature [^{O}C]	Maximum temperature [^{O}C]
3 mm	32.91	34.35	28.29	29.26
7 mm	38.27	40.64	31.06	33.94
11mm	40.01	43.11	36.26	39.54

In the previous studies, many of the researchers had considered dental drill guide culpable for rise in bone temperature during dental implant surgery and explored many options to control it. Misir et al. [9] conducted an in vitro study to determine the influence of surgical drill guide and drilling system on bone temperature and realized that implant site preparation using surgical guides results in added heat generation as compared to classical site preparation no matter which type of irrigation was used.
Researchers had tried to control the bone temperature by many means like changing the drill type, drill design, using different irrigation techniques (external / internal), using both the irrigation techniques simultaneously, using low temperature irrigant, changing the drilling parameters (speed, feed, drilling pressure etc) but no one has explored the idea of making changes in the drill guide itself. Gehrke et al. [10] studied the effects of various irrigation methods (external, internal and double irrigation) on bone temperature but did not find considerable advantages of using internal irrigation over the classic external irrigation. However, it was found that double irrigation approach delivers marginally improved outcomes.
Experimental comparison of drill guides shows encouraging results. Average temperature difference of 4.62 ^{O}C, 7.21 ^{O}C and 3.75 ^{O}C at the depth of 3, 7 and 11 mm were found. From the results, it is evident that the new drill guide is capable of achieving the set target. Temperature was reduced at all the locations, which means that flow of irrigant had increased at the surface as well as inside the cavity.

4. Conclusion

In the improved drill guide, irrigation channels provided in the bush / sleeve have resulted in efficient irrigant delivery. The small clearance between internal diameter of bush and drill no longer acts as a barrier for the supply of irrigant and thus improves fluid flow and its contact with drill and bone. The required amount of irrigant can now reach the drilling site without hampering its primary purpose i.e. placement accuracy. The dental drill guide with onsite irrigation delivery system helps in controlling the bone temperature during osteotomy site preparation because of the presence of local cooling system in the bush. This new drill guide with robust irrigation delivery system is expected to cost about 10 % more than the product available in the market.

5. References

[1] V. Kalidindi: *Optimization of Drill Design and Coolant Systems during Dental Implant Surgery,* Thesis, University of Kentucky Master's Theses (2004), p. 314.

[2] O.S. Gbadebo, F.B. Lawal, A.O. Sulaiman, and D.M. Ajayi: *Dental implant as an option for tooth replacement: The awareness of patients at a tertiary hospital in a developing country,* Contemporary Clinical Dentistry, Vol. 5, Issue 3 (2014), p. 302-306.

[3] K. Shigli, M. Hebbal and G.S. Angadi: *Attitudes towards Replacement of Teeth among Patients at the Institute of Dental Sciences, Belgaum, India.* Journal of Dental Education, Vol. 71, No. 11 (2007) p. 1467-1475

[4] J.B. park, Y.K. Kim, in: *Biomedical Engineering Fundamentals*, edited by J.D. Bronzino, third edition, Chapter 38, CRC press Taylor & Francis group (2008).

[5] A. Visetvitsakul, T. Kusalanukhan, C. Sinthanayouthina, and W. Tharanon: *Computerized design for implant position guide from any commercial implant planning software*, in 5th International Conference on Electrical Engineering/ Electronics, Computer, Telecommunications and in Krabi, Thailand, 2008. IEEE pp. 73-76.

[6] Prima Surgical Manual, Keystone Dental Information on www.keystonedental.com

[7] B.C. Sener, G. Dergin, B. Gursoy, E. Kelesoglu and I. Slih: *Effects of irrigation temperature on heat control in vitro at different drilling depths*, Clinical Oral Implants, Res. 20 (2009), p. 294-298.

[8] J. Lee, O.B. Ozdoganlar and Y. Rabin: *An experimental investigation on thermal exposure during bone drilling,* Medical Engineering & Physics, Vol. 34, Issue. 10 (2012), p. 1510-1520.

[9] F. Misir, M. Sumer, M. Yenisey and E. Ergioglu: *Effect of surgical drill guide on heat generated from implant drilling*, Journal of Oral and Maxillofacial Surgery, Vol. 67 (2009), p. 2663-2668.

[10] S. A. Gehrke, H.L. Neto, F.E.C. Mardegan: *Investigation of the effect of movement and irrigation systems on temperature in the conventional drilling of cortical bone,* British Journal of Oral and Maxillofacial Surgery. Vol. 51, Issue. 8 (2013), p. 953–957.

IOP Conf. Series: Materials Science and Engineering **422** (2018) 012020 doi:10.1088/1757-899X/422/1/012020

Product Innovation by Means of Smart Materials: The Case of Daxorol

Shpresa Syla, Hyrije Abazi-Alili and Florin Abazi

South East European University, Ilindenska 335, 1200 Tetovo, Republic of Macedonia

E-mail: h.abazi@seeu.edu.mk

Abstract. This research aims at investigating the role and function of smart materials as a flexible approach in developing innovative products, and their impact on companies' performance. The paper first elaborates smart materials as a new trend in technology, by further identifying types of smart materials. Supplementary, some benefits of using smart materials in companies are named, especially for a case study on using smart paint and coating for producing innovative product. The concluding part responds to the question: How will product innovation by means of smart materials improve companies' performance?

1. Introduction

Being different is considered as an advantage in the modern millennium. Nowadays, firms perform under a greater pressure than before because of competition that they face in the market. Usually, they offer the same or similar product or service, or they are under the pressure of the customers who expect and require more and more from the product they consume or service that they use.

In order to face with the new conditions and situations, firms are forced to continuously search for new ways of acting; they should continuously bring to the market innovations, namely to offer new products and services, or enhancing the existing ones.

Considering the issue of firm performance, it has been central in strategy research for decades and encompasses issues raised in the field, such as, why firms differ, how they behave, how they choose strategies and how they are managed [1]. A central premise of the resource-based view (RBV) is that firms compete on the basis of their resources and capabilities [2].

By a resource is meant "anything which could be thought of as a strength or weakness of a given firm, for example: brand names, in-house knowledge of technology, employment of skilled personnel, trade contacts, machinery, efficient procedures, capital, etc. [3].

This paper aims at identifying the advantages of innovative product by means of smart materials, in terms of competitive advantage and firm performance. In order to do so, the paper employs case study analysis, i.e. a factory that produces colors and coating in Macedonia, named Daxorol, is under the loop of analysis.

The structure of the paper is as follows. Fist, the empirical evidence of the impact of product innovation on firm performance is treated. The link between product innovation and smart materials is explained, before introducing the characteristics of the case study. Afterwards the paper presents some concluding remarks.

2. Product Innovation by means of Smart Materials

Smart materials are known as well as the materials that respond to environmental stimulation with particular changes in some variables. Due to this, they are also called as responsive materials. In response to their environment these materials can change their mechanical properties: shape, stiffness, and viscosity (or electromagnetic properties) in a foreseeable or controllable manner. The term "Smart materials" is applicable to materials and systems that can responsively react to change interior environments through material properties or material synthesis.

There are many groups of smart materials, each exhibiting particular properties which can be harnessed in a variety of high-tech and everyday applications. Smart materials have been around for many years and they have found a large number of applications. As such, smart material technologies are the key to 21st-century competitive advantage. Thus, various companies that work with building materials can significantly increase levels of functionality by using smart technologies and smart materials. As a result, the building methods and materials will be changed by the importance of smart materials.

Painting and coatings are ancient techniques for changing or improving the characteristics or performance of a material. The development of smart paints and coatings give these old approaches new capabilities. Smart paints and coatings can be generally classified into:

- high-performance materials,
- property-changing materials and
- energy-exchanging materials.

The pigments may be insoluble or soluble finely dispersed particles, the binder forms surface films. The liquid may be volatile or nonvolatile, but does not normally become part of the dried material. Coatings are a more generic term than paints and refer to a thicker layer. Many coatings are nonvolatile. These paints or coatings absorb energy from light, chemical or thermal sources and reemit photons to cause fluorescence, phosphorescence or afterglow lighting.

Before going to the analysis of the paints and coating factory as a case analysis, we will examine the relationship of product innovation and firm performance, with specific emphasis on the case of Macedonia.

3. Evidence on the Impact of Product Innovation on Firm Performance

Innovation is a process of transforming the new ideas, new knowledge into new products and services. Considering firms' innovative activities, it is unclear whether it leads to improved firm performance. One should expect that if the firm is innovative than it will respond to costumers needs, thus will be positively evaluated by the market as a good performing firm.

Different studies found the positive effects of innovation on firm-performance [4]. There are different driving forces that push companies towards innovation, such as R&D activities. Since R&D intensity is higher in EU member economies, they should be a leading example for other non-EU countries. Whenever there is a lack of knowledge to bring changes in the firm, we suggest that spending on the acquisition of new knowledge will have a positive effect on innovation activities and additionally on firm-performance [5].

Different studies found the positive effects of innovation on firm-performance [6,7]. It is also noted in vast literature that firms that engage in developing innovative products and services are positioned to compete more successfully through the development of new products and processes, increasing market share, return on investment (ROI), and overall firm success". Reference [8], applying metaanalyses techniques to aggregate prior empirical research on the innovation–performance relationship (42 empirical studies on 21,270 firms), have found that product innovation has a positive effect on the firm-performance.

So that a company undertakes innovation activities by means of smart materials, one should consider its technological intensity. The technological intensity of the industry in which a firm is operating is another variable that might have an impact on product innovation, and at the same time on firm performance.

A questionnaire based survey was conducted in 60 companies in Macedonia, in order to measure the scale in which they undertake product and process innovation. We have followed the Eurostat definition for the classification of components of the manufacturing industry on the basis of their technological

intensity, using NACE at 2-digit level as the basis of classification, grouping different activities into high-technology, medium high-technology, medium low-technology and low-technology industries.
Labor productivity is a measure for the performance of the firm, and it is cross-tabulated with technological intensity and innovativeness for identifying the current situation in Macedonia.
Fig. 1 shows the relationship between product innovation of firms in different technological classifications and their labor productivity.

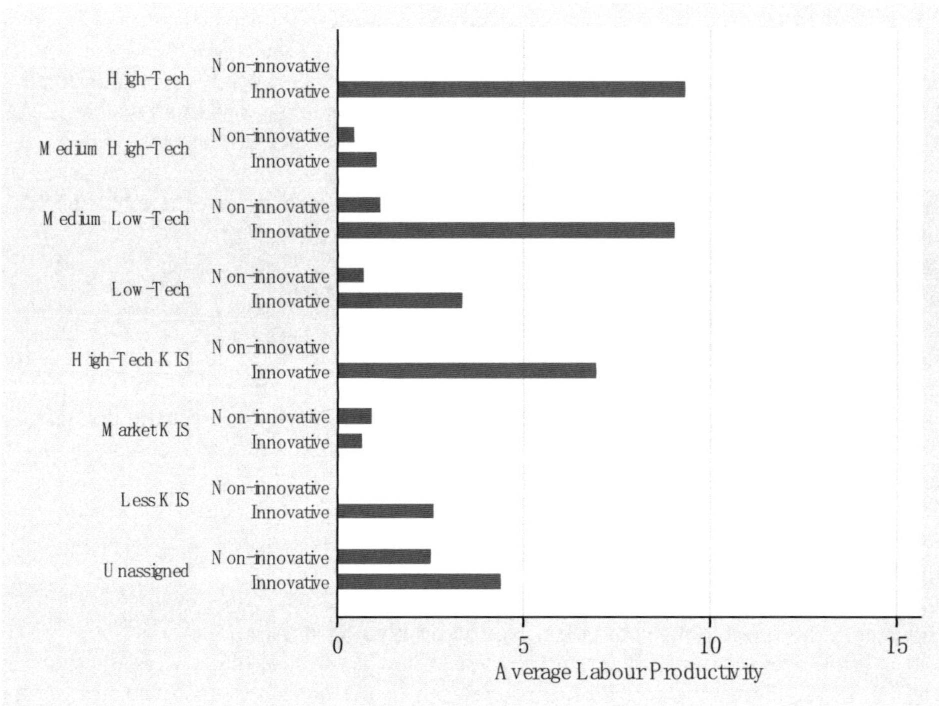

Figure 1. Labour productivity and innovativeness by technological intensity of firms (in million denar per worker).

Firms in the group of high-technology manufacturing or high knowledge intensive services seem to perform better than other firms. Also, innovative firms seem to be performing better than non-innovative firms across the sample. Interestingly, there are no non-innovative firms in the high-technology and high-tech KIS industry.

4. Case Study: Daxorol

In this paper, Daxorol will be examined as a case study of company that has introduced new product to become more competitive in the market. Daxorol is a factory that produces colors and coating in Macedonia. It is characteristic for this company that it employs advanced contemporary technology for production. It uses Daxomix to produce 1500 shades of colors, that can be applied on interior and facades.
In order to satisfy costumers' requirements and difficulties in choosing colors, Daxorol employs machinery to come up with innovative product. Since, the difficulty in choosing color by costumers is because they cannot imagine the effect of the change on the wall, Daxorol has found solution for them. With the Dinova color designer, they have the opportunity to try by their own on the first on-site appointment with a coherent color scheme. Furthermore, the customers can show even more competence with the new color measuring device Colorcatch Nano. With this innovation, it is possible to measure color tones from a size of 0.4 mm.

4.1. Dinova Colordesigner - the Application for Photo-Realistic Color Design Axomix
The application Dinova Colordesigner has the possibility to create color designs, that can be presented to the customers. The color designer is suitable for smartphones and tablets and is distinguished by its simple and intuitive handling. This application has many advantages that can be stated as follows: (i) Photorealistic design including background structure, light and shadow effect; (ii) Design possible in a few minutes and on site; (iii) Easy, intuitive handling; (iv) Very low cost; (v) Further strengthen your consulting competence.
4.2. Colorcatch Nano – Shade Measurement Has never Been More Accurate
With the color meter Colorcatch Nano one can save time and money. It incorporates a revolutionary technology to accurately measure the shades of smooth, textured or patterned surfaces. It reads over 50,000 pixels per measurement. The advantages of the Colorcatch Nano are as follows: (i) Pixel-accurate measurements possible with up to 5 dominant colors per measurement; (ii) Optimal illumination of all structures thanks to 9 integrated D65 LEDs (daylight); (iii) Lens with 6x magnification; (iv) Disturbances such as shadows, dust or dirt particles are calculated by algorithm; (v) Handy format, low weight and high mobility.

Figure 2. Dinova Colordesigner and Colorcatch Nano

The synergy of the abovementioned tools, makes the product of Daxorol more innovative and competitive in the market, and at the same time satisfies the requirements of the customers. Thus, what the company gets by this "Strong duo" is optimal interaction of Dinova Colordesigner and Colorcatch Nano, that results in more competence.
4.3. DaxoMix - the Professional Sound System from Daxorol
It is characteristic for this factory the precision and high quality innovative product as an output of DaxoMix. They cooperate exclusively with the best and most experienced color mixing plants manufacturers. The aim of the factory is that every DinoMix tweeter fulfills the high expectations of professional processors. The advancement in technology has brought to distinguished product that offers many advantages.
Daxorol is successful and high performing company as a result of product innovation.

5. Conclusion
This paper has set an accent on the importance of product innovation, especially by making use of the smart materials. It investigated (i) the concept of product innovation by means of smart materials and (ii) the impact of product innovation on the firm's performance, especially for the case of Macedonia.
The statistical overview of the Macedonian sample shows that technological intensity and innovative companies, have better performance measured by labor productivity.
Finally, some important findings can be identified from the case study analysis of this paper. Firstly, Daxorol incorporates new machinery and technology to satisfy customers' needs, and the result is improved performance. Secondly, being concerned with the advancement of the materials used, makes these company more professional, by coming up with precisely identified colors and materials. Finally,

the relation between product innovation and labor productivity in Macedonia has shown to be positively related. We come to this conclusion, as no matter the crisis that the country has faced lately, Daxorol has showed better performing results.

6. References

[1] M. E. Porter, Towards a dynamic theory of strategy. Strategic Management Journal, vol. 12, (1991), p. 95-117.

[2] M.A. Peteraf and M. E. Bergen, Scanning Dynamic Competitive Landscapes: A Market-Based and Resource-Based Framework. Strategic Management Journal, vol. 24, (2003), p. 1027–1041.

[3] B. Wernerfelt, A Resource-based View of the Firm. Strategic Management Journal, vol. 5, (1984), p. 171-180.

[4] H. Abazi-Alili, I. Hashi, and F. Abazi, The impact of innovation activities on firm performance: Evidence from transition economies, South East European Journal of Economics and Business, vol. 12, (2017), p. 7-17. DOI: 10.1515/jeb-2017-0012

[5] V.Ramadani, H. Abazi-Alili, L.P. Dana, G. Rexhepi, and S. Ibraimi, The impact of knowledge spillovers and innovation on firm-performance: findings from the Balkans countries. International Entrepreneurship and Management Journal, vol. 13 (1), (2017), p. 299–325.

[6] C. Fernades, J. Ferreira, and M. Raposo, Drivers to firm innovation and their effects on performance: an international comparison. International Entrepreneurship and Management Journal, vol. 9(4), (2013), pp. 557–580.

[7] V. Ratten, Behavioral intentions to adopt technological innovations: the role of trust, innovation and performance. International Journal of Enterprise Information Systems, 10(3), (2014), p. 1–13.

[8] N. Rosenbusch, J. Brinckmann and A. Bausch, Is innovation always beneficial? A meta-analysis of the relationship between innovation and performance in SMEs, Journal of Business Venturing, vol. 26, (2011), p. 441–457.

IOP Conf. Series: Materials Science and Engineering **422** (2018) 012021 doi:10.1088/1757-899X/422/1/012021

Investigation on the Precision Loss of Ball Screw Considering the Full Ball Load Distribution

Jiajia Zhao[1,a], Mingxing Lin[1,b], Xianchun Song[2,c], Hongkui Jiang[2,d], Xiaojian Wu[1,e]

[1]National Demonstration Center for Experimental Mechanical Engineering Education (Shandong University), Key Laboratory of High-efficiency and Clean Mechanical Manufacture of Ministry of Eduction, School of Mechanical Engineering, Shandong University, Jinan, Shandong, 250061, China
[2]Key Laboratory of Mechanical Engineering Innovation Technology, School of Mechanical and Electrical Engineering, Shandong Jianzhu University, Jinan, Shandong, 250101, China

E-mail: [a]zhaojia0922@126.com, [b]mxlin@sdu.edu.cn, [c]songxch@sdu.edu.cn, [d]jhk_2001@sdjzu.edu.cn, [e]xjwu@sdu.edu.cn

Abstract: Ball screws are the driving functional components most frequently used for the machining equipments. However, the precision positioning error of ball screws decide directly machining accuracy of machine tools. This paper aims to investigate the precision loss of ball screw considering full ball load distribution. The raceway wear model is built to predict the precision loss of ball screw in different structural parameters and operating parameters. Meanwhile, the precision loss rate is obtained and analyzed during the whole running life.

1. Introduction

With the development of high speed cutting, the dynamic performance of ball screw is getting higher and higher. As in many mechanisms, the wear in ball screws complicates the control design progress and reduces the positioning accuracy significantly. Lin [1] investigated the quasi-static contact characteristics of ball screw. Yoshida T [2] established a new finite element model to study the uneven distribution of stress. Mei [3] studied the influence of geometrical random error on the load distribution of ball screw. C. Braccesi [4] presented a general elastic plastic approach to impact analysis for stress state limit evaluation in ball screw bearings return system. Song [5] established a new online measuring system of ball screw movement precision. With the wear of ball screw, the accuracy of ball screw is reduced rapidly, which directly affects the life and the reliability of machine tool. A new model is built to predict the precision loss rate in different structural parameters and operating parameters.

2. Theoretical analysis

2.1 The load distribution of the full ball

The contact deformation δ between the ball and the nut or the screw are expressed as follows:

$$\delta_{ji} = C_j \cdot Q_i^{\frac{2}{3}}. \tag{1}$$

δ_{ji} (j=s *or* n) are the contact deformation between the ith ball and the screw or the nut respectively, C_j (j=s *or* n) are the contact stiffness of the screw or the nut respectively, Q_i is the normal load between the ith ball and the raceway.The axial deformation of the screw or the nut can be obtained respectively.

$$\varepsilon_{ji} = \frac{F_{i-1,i} \cdot \Delta L}{E \cdot A_j} \;,\; \Delta L = l_p / N'. \tag{2}$$

ε_{ji} (j=s *or* n) are the axial deformation of the screw segment or the nut segment between the i-1 ball and the i ball, E is the elastic modulus of the screw and the nut, A_j (j=s *or* n) are the cross-sectional area of the screw or the nut in the contact point, ΔL is the axial distance between two adjacent ball, $F_{i-1,i}$ is the axial load of the screw segment between the i-1th ball and the ith ball. l_P is the lead of the screw, N' is the number of balls in a single lead.Where the axial force relation of the screw along the axis is shown in Fig.1.

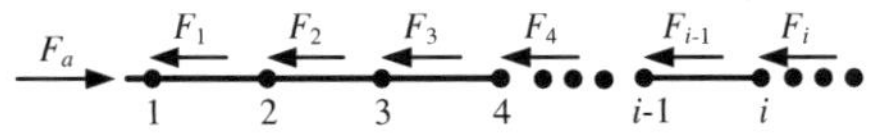

Figure 1. The axial force diagram of the screw axis

Where F_a is the axial load, F_i is the axial force between the ith ball and the raceway.
The relation of the axial force at each contact point can be obtained as follows:

$$F_{i-1,i} = F_a - \sum_{j=1}^{i-1} F_j = \sum_{j=1}^{N} F_j - \sum_{j=1}^{i-1} F_j = \sum_{j=i}^{N} F_j \;,\; F_i = Q_i \cdot \sin\alpha \cdot \cos\lambda. \tag{3}$$

Where N is the number of ball, α is the contact angle between the ball and the raceway, λ is the nominal helix angle of the screw. The deformation relation of ball screw between the i-1 ball and the i ball is shown in Fig.2. δ_{jai-1} and δ_{jai} are the axial component of the Hertz contact deformation of the i-1th ball and the ith ball with the nut raceway or the screw raceway respectively (j=s *or* n).The deformation coordination relationship can be obtained as follows:

$$\varepsilon_{si} + \varepsilon_{ni} = (\delta_{sai-1} - \delta_{sai}) + (\delta_{nai-1} - \delta_{nai}). \tag{4}$$

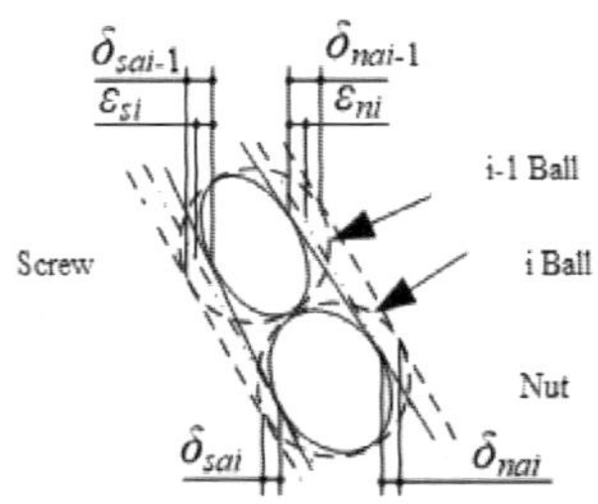

Figure 2. The deformation diagram of ball screw.

The axial component of the Hertz contact deformation between the ball and the nut raceway or the screw raceway can be obtained as follows [6]

$$\delta_a = \delta \cdot \sin\alpha \cdot \cos\lambda. \tag{5}$$

By substituting (1), (2), (3) and (5) into (4), the iterative formula of full ball load distribution can be obtained as follows:

$$\begin{cases} Q_i^{\frac{2}{3}} - Q_{i-1}^{\frac{2}{3}} + \frac{\Delta L}{E} \cdot (\frac{1}{A_s} + \frac{1}{A_n}) / (C_s + C_n) \sum_{j=i}^{N} Q_j = 0 \\ F_a = \sum_{i=1}^{N} Q_i \cdot \sin\alpha \cdot \cos\lambda \end{cases} \tag{6}$$

2.2 The wear model of the ball-screw contacts

Based on the Archard theory, the precision loss of the ball screw in the wear can be expressed as follows [7]

$$W_{ji} = K \frac{Q_i L_j}{H}. \tag{7}$$

Where W_{ji} (j=*s or n*) is the amount of wear volume, Q_i is the normal pressure of *i*th ball contact point, L_j is the relative sliding distance between ball and raceway at contact points, K is the coefficient of wear, H is the hardness of raceway.The contact area of ball with the screw or the nut can be written as follows:

$$A_{ji} = \pi a_{ji} \cdot b_{ji}. \tag{8}$$

Where a_{ji} and b_{ji} are the semi-major axis and semi-minor axis on the contact ellipse respectively.

Considerating to the period of a ball passing through the contact point on the raceways of the screw or the nut, the coefficient f_s, f_n is introduced respectively. Thus, the total sliding distance of a contact point on the raceways of the screw or the nut can be obtained as follows:

$$L_s = f_s \cdot \left|\vec{V_s}\right| t = \frac{b_s \sin\lambda}{l_s} \cdot \left|\vec{V_s}\right| t, \quad L_n = f_n \cdot \left|\vec{V_n}\right| t = \frac{b_n \sin\lambda}{\mathrm{m} l_p} \cdot \left|\vec{V_n}\right| t. \tag{9}$$

Where t is the running time, l_s is the effective travel of the screw, m is the number of turns in one chain and V_j (j=*s or n*) are the ball's slipping velocity relative to the raceway at the contact point between a ball and the screw or the nut respectively.

By combining (7), (8), (9) and (10), the wear depth of the raceway can be obtained as follows:

$$\Delta\delta_a = \sum_{i=1}^{N} \frac{W_{si}}{A_{si}} + \sum_{i=1}^{N} \frac{W_{ni}}{A_{ni}} = \sum_{i=1}^{N} \left(\frac{K_s Q_i \sin\lambda \left|\vec{V_s}\right|}{\pi H a_s l_s} + \frac{K_n Q_n \sin\lambda \left|\vec{V_n}\right|}{\pi H a_n \mathrm{m} L_P} \right) \cdot t = \delta_{wt} \cdot t. \tag{10}$$

Where δ_{wt} is the precision loss rate of ball screw.

Table 1. Parameters used in the simulation test

Parameters	Value	Unit
Nominal diameter	40	mm
Ball diameter	4.763	mm
Lead angle	3.64	°
Lead	8	mm
Rows ×Turns	3x1	
Contact angle	45	°
Poisson's ratio	0.3	
Young's modulus	205	GPa

3. Results and discussion

It is worth mentioning that the running-in process is ignored in this paper to study the precision loss of ball screw because of wear. The precision loss rate of ball screw versus axial load in different rotational speeds are shown in Fig.3. The relation of precision loss rate versus the axial load and rotational speed can be described as syntagmatic spatial plane. The precision loss rate is increased with an increasing axial load and rotational speed. When the axial load is increased, the normal force and the contact ellipse area between the ball and the raceway increase with the increasing axial load. However, the increasing rate of contact normal force is larger than the increasing rate of contact deformation area. It is obvious that the precision loss rate is proportional to the normal force and inversely proportional to the contact area. Thus, for a given rotational speed, the precision loss rate increases with increasing axial load. For a given axial load, the precision loss rate increases with increasing rotational speed.

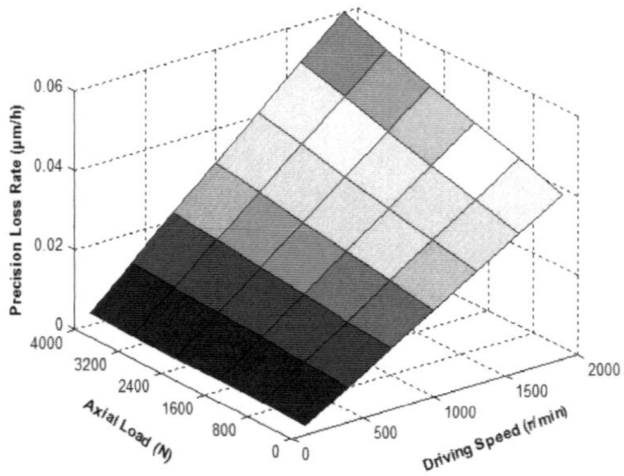

Figure3. Precision loss rate of the raceway versus axial load at different rotational speeds

Fig.4 show the precision loss rate of ball screw in different structure parameters. The precision loss rate of ball screw decreases with the increase of the contact angle. Thus, the greater contact angle can reduces the precision loss of ball screw. In addition, The precision loss rate of ball screw increases with the increase of the helix angle. Thus, the precision loss of large lead ball screw is serious, and properly reducing helix angle can decrease the precision loss of ball screw. The precision loss rate of ball screw increases with the increase of the raceway radius. Whlie, the precision loss rate of ball screw decreases with the increase of ball diameter. The turning point is the situation that the ball radius is equal to the raceway radius. Therefore, It's obvious that reasonable structure parameters can improve the precision preservation of ball screw.

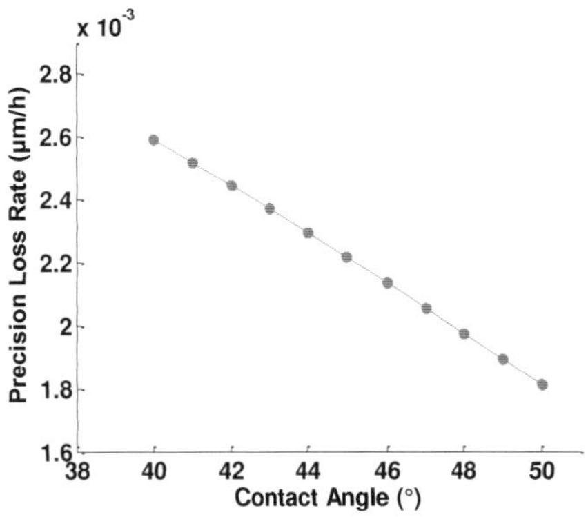

(a)

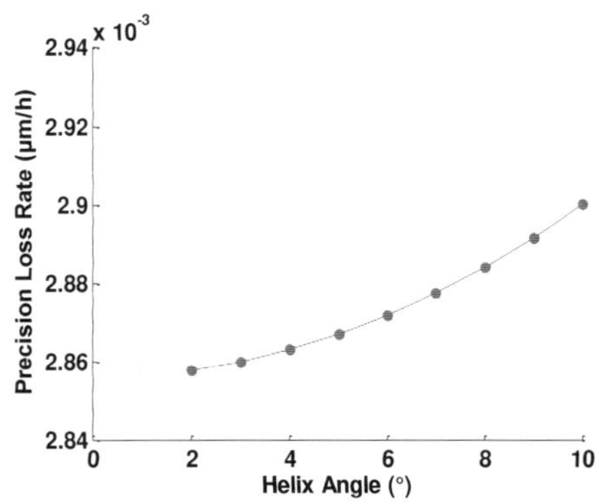

(b)

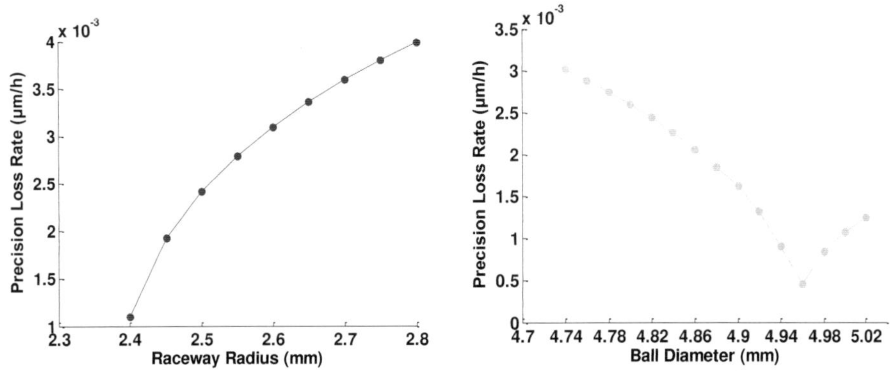

Figure 4. Precision loss rate of the raceway in different structure parameters, (a) contact angle, (b) helix angle, (c) raceway radius and (d) ball diameter.

4. Conclusion

High-precision machine tools can not be separated from high-precision functional components. As a key functional component, the accuracy of ball screw directly affects the positioning precision of machine tools. The wear model is built to predict the precision loss of the ball screw in different structural parameters and operating parameters. Meanwhile, the precision loss rate is obtained and analyzed during the whole running life.

5. Acknowledgement

This work was supported by the National Natural Science Foundation of China (51475267).

6. References

[1] M. C. Lin, Design and mechanics of the ball screw mechanism. *The University of Wisconsin-Madison*, (1989).
[2] Yoshida T., Tozaki Y., Study on load distribution and motion of ball screws. *Journal of Japanese Society of tribologists*, Vol. 8 (2003), p.659-666.
[3] X.S. Mei, Study on the load distribution of ball screw with error. *Mechanism and Machine Theory*, Vol. 38 (2003), p.1257-1269.
[4] C. Braccesi, L. Landi, A general elastic plastic approach to impact analysis for stress state limit evaluation in ball screw bearings return system. *International Journal of Impact Engineering*, Vol. 34 (2007), p.1272-1285.
[5] Xianchun Song, Yuankai Lu, Screw motion precision on-line measurement system design and experimental analysis. *Journal of Shandong Jianzhu University*, Vol. 35 (2017), p.409-413.
[6] Hertz H., On the contact of elastic solids, *Journal Für die Reine und Angewandte Mathematik*, Vol. 92 (1881), p.156-171.
[7] Zhou C.G., Ou Y. and Feng H.T., Investigation of the precision loss for ball screw raceway based on the modified Archard theory, *Industrial Lubrication and Tribology*, Vol.2 (2017), p.166-173.

Parameterized Modeling and Material Intensity Simulation of Excavator Stick

Jun Zhang[1] , Shuli Sun

School of Engineering, Zhejiang University City College, Hangzhou 310015, China

[1]E-mail: zucczj@163.com

Abstract. In order to shorten the design cycle of excavators and to meet the needs of users constantly updating key parameters in their designs, a reusable parametric design model of excavator sticks has been developed. Based on the WAVE function of Siemens NX, a parameterized model of large excavator stick is constructed, and a codeless reusable system is constructed by using NX / PTS powerful interactive operation function. At last, the reliability of the model is verified by the strength simulation .The entire system interface is intuitive and easy to operate, which greatly shortens the design cycle of the excavator stick and improves the design efficiency of the stick.

1. Introduction

Excavator is a kind of construction machine used for excavating soil. It is mainly used for excavation of trench in road construction, dredging of river in hydraulic engineering and ore excavation in mining engineering [1]. According to statistics, about 60% of the earthwork volume of the construction of the project is achieved by excavators. According to the development of construction machinery in recent years, the excavator has become one of the most important construction machinery in the construction. Excavator main components are sticks, buckets, boom, rotary platform, power transmission and walking agencies [2]. The stick is an important component that connects the boom to the bucket, and its structure is complicated, which often takes a lot of manpower and money in design. In order to complete the complex structure of product design in a relatively short period of time and meet the rapidly changing market demand, it is necessary to realize the rapid design of the parametric model of excavator main components.

NX is an interactive CAD/CAE/CAM system. WAVE(What-if Alternative Value Engineering) is a powerful module in NX and the top-down modelling approach is WAVE's guiding principle that makes it possible to design a top-down association between product components. PTS (Product Template Studio) is a functional module in NX that implements code-free human-computer interaction design, which makes it possible to design object-oriented visualization parameters. Based on the traditional design experience of excavator, this paper constructs a parametric reusable design model of excavator stick by using WAVE and PTS module functions, and uses ANSYS and ADAMS software to simulate the strength of the stick to verify the reliability of the model [3]. This model greatly shortens the design cycle of the excavator stick and improves the re-design efficiency of the stick.

2. Parametric Model of Stick

2.1. Siemens NX

Siemens NX is an all-in-one product engineering solution that drives product innovation through process changes. WAVE is a NX technology that enables the modelling of relationships among product components. It effectively controls the assembly design and part design of complex products to ensure the interconnection of the entire assembly and components. PTS is a reusable module of NX that significantly improves the re-use of design knowledge by creating product templates, and can also create intuitive and interactive interface for model parameter called, in order to achieve the related control parameters.

2.2. The functional composition of the stick parametric model

Stick is an important part of the excavator, mainly by the root bracket, front bracket, the middle bracket and the connecting plate, as shown in Figure 1. The parametric model of the stick must be designed to be easy to use and will not cause confusion in the control of the parameters and to achieve the following functions:

(1) Open the template browser through "Edit Reusable Components" in the NX Assembly Navigator. The key dimension parameters in the related components can be modified, and the model can be updated with the parameter changes.

(2) When the parameters of the stick top model are modified, the bottom sub-part models will also be updated; when the parameters of the stick bottom model are modified, the same level model associated with it will change, and the top model of the stick will also be updated.

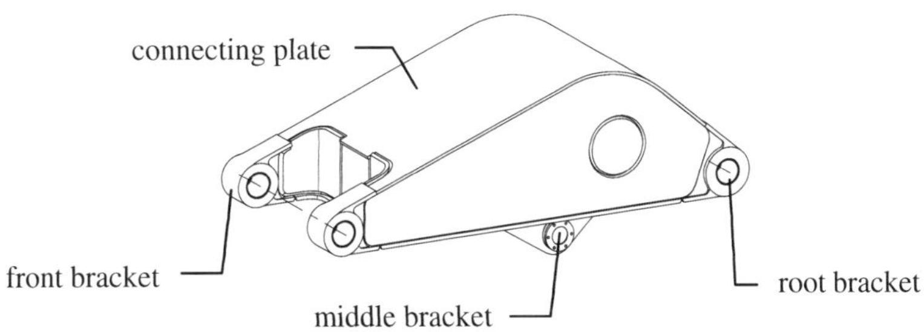

Figure 1. Stick assembly.

3. Construction of Parametric Model Based on WAVE

Parametric modelling is the core of CAD technology. At present, most CAD software has the function of parametric modelling, and WAVE is the top-down parametric assembly modelling function of NX. The key to parametric modelling is using parameters, formulas, links, features, etc. to drive and change the model. It is the premise of using PTS to make product templates and the key to realizing knowledge reuse.

3.1. Determination of overall control parameters for stick

The overall control parameters mainly refer to the parameters that affect the function and structure of the product. When using NX / WAVE function to parameterize the excavator stick, we first need to analyse the structural design demand of the stick by relying on the accumulated experience of the traditional design of excavator stick. Then according to the characteristics of the structure of the stick, the root of the stick is set as the origin of coordinates, and then the overall control parameters are obtained as follows: L1 (horizontal coordinate of the front bracket hole), L2 (horizontal coordinate of the middle bracket hole), H (vertical coordinate of the middle bracket hole). As shown in Figure 2.

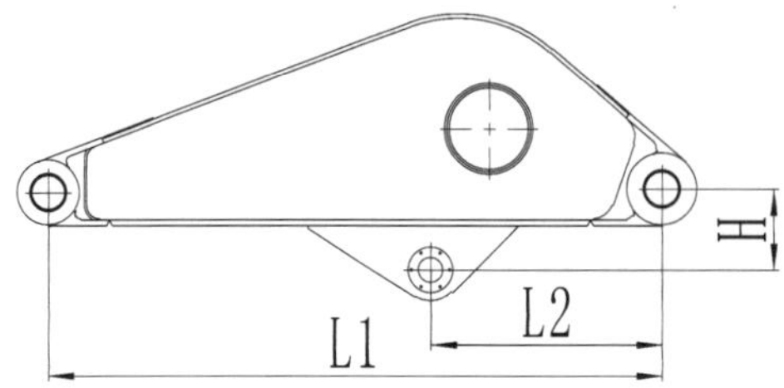

Figure 2. Overall control parameters of stick.

3.2. Determining the associated structure of stick

The composition of the excavator stick is more complicated. In order to show the clear associative design structure of the stick and facilitate the operation and management of the designer, we define the control structure between the stick components and the parts, according to the symmetrical structure of the stick and the interrelationship between sketches and benchmarks of each component, as shown in Figure 3.

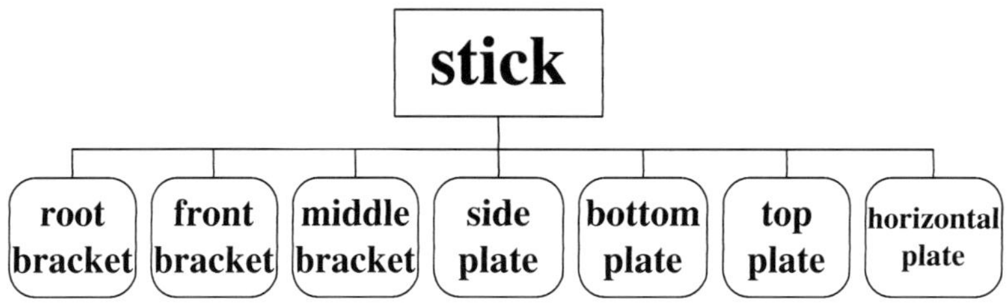

Figure 3. Associated structure of stick.

3.3. Construction of stick parametric model based on WAVE

The parametric modeling of the excavator stick adopts the top-down design method. The stick model is constructed by two layers in sequence according to the associated structure diagram:

3.3.1.Top components. Firstly, we use NX modeling module to create a 3D empty model of excavator stick and name it “stick" as the top component of the entire assembly model. Then use the expressions, datum, and sketches to create three overall control parameters for the stick in the empty model. As shown in Figure 4.

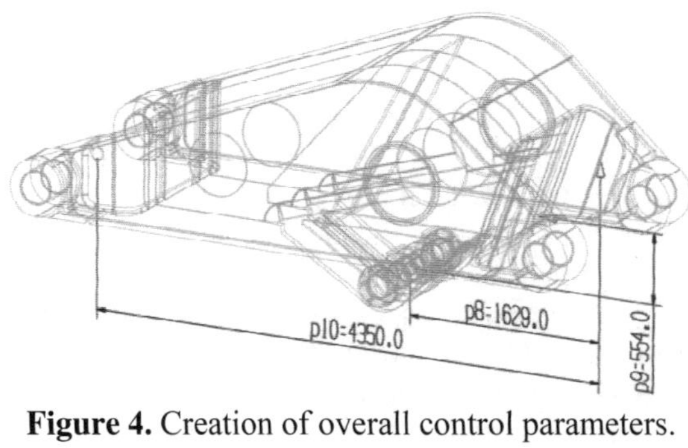

Figure 4. Creation of overall control parameters.

3.3.2. Bottom components. Firstly, we use NX/WAVE "New Level" function to create each subcomponent of the stick as the bottom components of the assembly model and named them " root bracket ", " front bracket ", " middle bracket ", " horizontal plate ", " bottom plate ", " top plate ", " side plate ", and then use the WAVE Geometry Linker to copy the geometry features of the top components to the newly created bottom components. We use this as a base plus a new sketch to create the bottom geometry. Taking the root bracket as an example, the parametric model is created as shown in Figure 5.

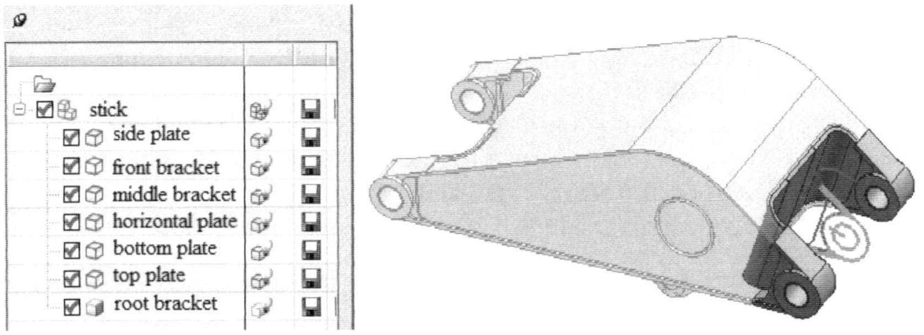

Figure 5. Construction of parametric model of root bracket

Need to pay attention to: Sketching process must be in line with the specifications to ensure that all constraints and cannot have virtual constraints or over constraints. Full constraints may have a variety of design schemes; the design experience needs to be based on a more reasonable selection of all constraints in order to meet the design requirements. The bottom sub-components may also need to be related to each other through the WAVE, in order to achieve the purpose of updating simultaneously.

3.4. The creation of stick 2d drawings

In the actual design process of the excavator stick, after the 3d model of the part is created, 2d drawings conforming to the corresponding standards and specifications are also to be designed and produced. We also need to mark the size and tolerance, display style and attribute filling operation, and finally to save as the 2d drawings of the stick.

4. Construction of Interactive Dialogs Based on PTS

PTS module is NX user-oriented, code-free parameter dialog design tools, the establishment of the expression, sketch data, etc. simply by dragging can be passed to the PTS, and generate user interface with display, data input and other functions. The user interface can be directly embedded into the NX assembly model file, and the user can modify the model parameters through a simple interface operation.

PTS dialog box design is divided into the following steps: Start PTS and import the parametric model of stick, and double-click the model thumbnail into the user interface. In the work area, add items such as type block, group, separator and bitmap etc. In the model operation interface, we drag and drop the relevant dimensions and expressions established in the parametric modelling into the corresponding groups, and add the created 2D drawing to the PTS to establish the connection between the model and the 2D drawing. The design dialog box for overall control parameters and root bracket dimensions are shown in Figure 6.

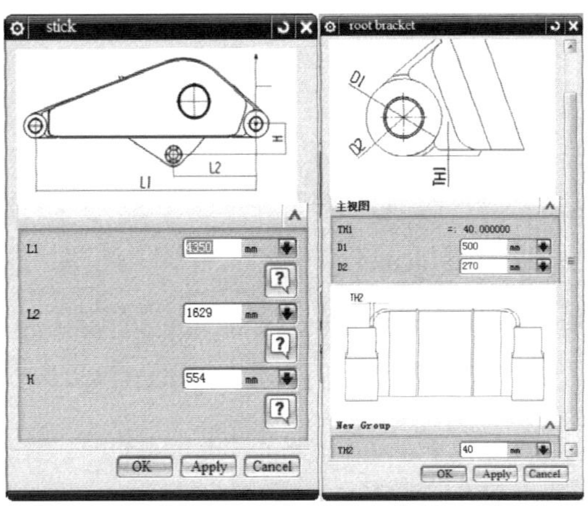

Figure 6. The dialog box of stick & root bracket

5. Operation of Stick Parametric Model

Start NX, open the "stick" .prt file, select the top-level component in the Assembly Navigator in the modeling environment and right-click on "Edit Reusable Components" to bring up the template browser and enter the parametric modeling dialog box. We can double-click to open the top and bottom components in the template browser, modify the appropriate dimension to achieve the model and 2d drawings associated updates. Taking L1 about overall control parameters of stick as an example, before and after the modification model comparison shown in Figure 7.

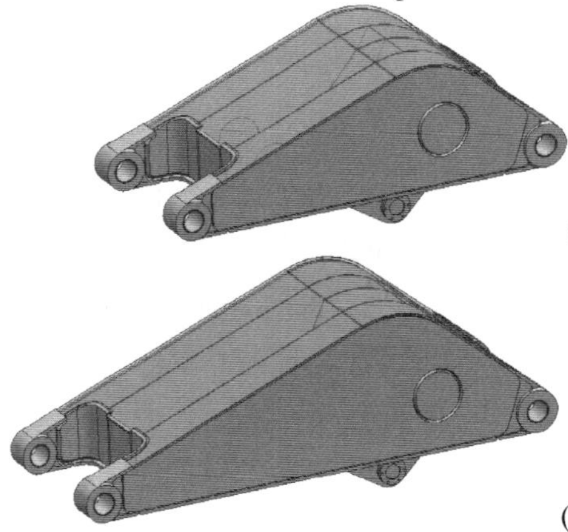

(a) L1=4350mm

(b) L1=5500mm

Figure 7. Comparison chart of L1 changes before and after.

6. Simulation of Stick Material Intensity

We use the finite element software ANSYS and multi-body dynamics software ADAMS together to achieve the strength simulation of the stick. First import the stick model into ANSYS, create an modal neutral file (.mnf) of stick by operation, then import this modal neutral file into ADAMS and assemble it with the bucket and boom, then apply the constraints and loads in the ADAMS to perform the dynamics simulation, and finally get the stress map of the stick, as shown in Figure 8, We can see that the maximum stress of the stick is 205MPa, which is within the range of the stress limit of steel 300Mpa, which meets the design requirements.

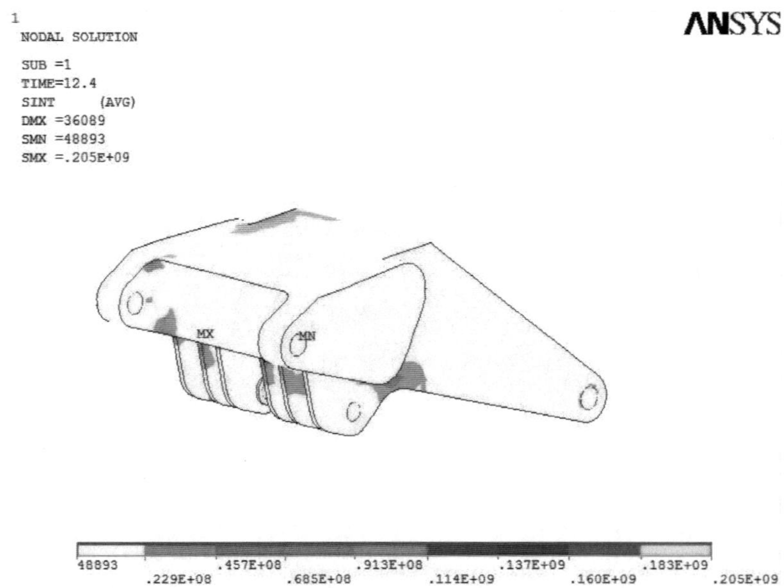

Figure 8. Stick stress diagram

7. Summary

In this paper, the parametric construction of the excavator stick model is realized by using the WAVE technology, parameter association and feature chaining. An interactive dialogue box of the stick is created by using PTS module. It realizes the function that modifying the size of the top model to drive the modification of the associated subcomponent model and modifying the size of the bottom model to drive the modification of the interrelated bottom model. Finally, the reliability of the model is verified by the structural strength simulation. The interactive dialog interface of the parametric model is simple, easy to operate, can greatly improve the re-design efficiency.

8. Acknowledgments

This research was supported by the Laboratory Research Project of Zhejiang Provincial.

9. References

[1] Huanyu Zhao, Guoqiang Wang, Haotong Wang, Qiushi Bi, Xuefei Li, Fatigue life analysis of crawler chain link of excavator. In Engineering Failure Analysis, 79(2017) 737-748.

[2] Faridaddin Vahdatikhaki, Amin Hammad, Dynamic equipment workspace generation for improving earthwork safety using real-time location system. In Advanced Engineering Informatics, 29(2015) 459-471.

[3] Shuxiang Guo, Yanlin He, Liwei Shi, Shaowu Pan, Rui Xiao, Kun Tang, Ping Guo, Modeling and experimental evaluation of an improved amphibious robot with compact structure. Robotics and Computer-Integrated Manufacturing, 51(2018) 37-52.

Overall Strength Analysis of Trailing Arm Torsion Beam Suspension System

Yun-Xia WANG [1], Jin-Ting NI[1], Ai-Guo Wang[2]
[1,2] Dept. of Automotive Eng., Anhui Technical College of Mechanical and Electrical Engineering, China

E-mail:ahjdwyx2015@126.com

Abstract. Previous studies focusing on analyzing components of automotive suspension system separately usually ignore the fact that each single component makes movement during operation resulting in simulation bias. This paper carries out a thorough analysis on the overall strength of a trailing arm torsion beam suspension system and compares the results with previous ones from analyzing single pieces. Consequently, significant improvements will be made in the design of suspension system.

1. Introduction

Trailing arm torsion beam suspension is a relatively new rear suspension system, to which two longitudinal arms with each attached to a separate wheel is welded. It is a commonly-seen rear suspension system in automobiles. It functions to pass through various forces and movements between the body and wheels, alleviate impact load passed by uneven roads, and control the movement of wheels as to ensure the ride comfort and handling stability. Therefore, the strength of the suspension system has a direct impact on the reliability and life span of vehicles [1].

The main components of the railing arm torsion beam suspension system include coil spring, shock absorber, trailing arms, shock absorber bracket, hub carrier bracket, trailing arm reinforcing plate and rear reinforcing plate. Previous studies on the strength of the suspension system only concern the strength check of components neglecting the fact that components make movements during operation, which leads to the result of miscalculation and inaccuracy [2-4].

This paper proposes to analyze the overall strength of the suspension system by use of the trailing arm torsion beam rear suspension system of a real car as a sample, aiming to establish constraint relationships between various components and model elastic units, such as coil spring, shock absorber, and bushing. Subsequent results will be compared with previous results on single components, based on which reasons explaining the differences will be concluded [5].

2. Finite Element Modeling and Load Cases Determined

(1) Finite Element Modeling. The trailing arm torsion beam rear suspension system of the sample mainly consists of coil spring, shock absorbers, rear beam, trailing arm, shock absorber bracket, hub carrier bracket, trailing arms reinforcing plate, and rear spar reinforcing plate. A geometric model is established in 3D software CATIA based on the design drawings, as shown in Figure 1. Then, analyzing material properties, defining connection properties between components and boundary conditions are completed in HYPERMESH software to generate finite element model, followed by calculation in ABAQUS software.

Real beam, trailing arm, shock absorber bracket, trailing arm reinforcing plate, rear reinforcing plate are divided by SHELL elements; welding spots are simulated by RIGID element; welding line is simulated by SHELL element with the same thickness; hub seat holder is divided by the second-order tetrahedral elements; spring members including spring and shock absorber are simulated by bush unit, while bush connection parts are given corresponding nonlinear elastic properties. Overall, 39352 elements and 42242 nodes, form the finite element model, as shown in Figure 2.

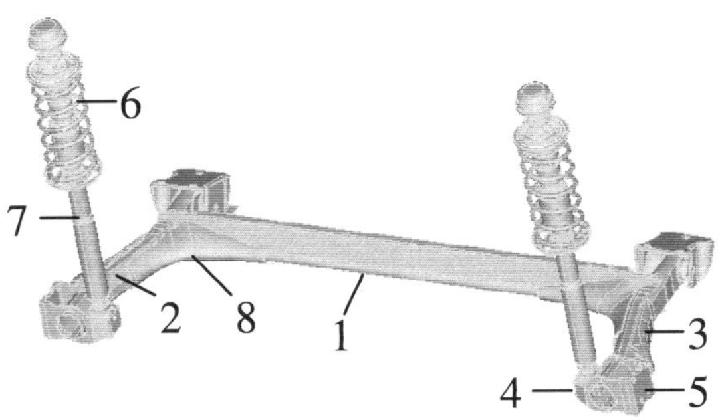

1-rear beam; 2-trailing arm; 3-trailing arm reinforcing plate; 4-shock absorber bracket; 5-hub carrier bracket; 6-coil spring; 7-shock absorber; 8-rear spar reinforcing plate

Figure 1. CAD model of a trailing arm torsion beam suspension

Figure 2. Finite element model of the railing arm torsion beam suspension system

(2) Load Case. Load born by wheels in motion is most likely random, but sometimes varies according to the load of vehicle, driver's operation, engine output torque, road roughness, surface adhesion coefficient and speed. It is, therefore, of significant importance to determine the typical operating conditions as the strength check based on mechanical characteristics of the suspension [1].

The forces of suspension system can be summarized under four typical working conditions: (1) The maximum vertical force condition, namely full-load vehicle on uneven roads withstands impact loads; (2) The maximum braking power condition, namely full-load vehicle hits emergency braking; (3) The maximum longitudinal force condition, namely full-load vehicle applies Linear Acceleration; (4) The maximum lateral force condition, namely full-load vehicle make turns. Calculated as follows:

(1) The maximum vertical force conditions, the maximum vertical force F_{Vm} of unilateral rear wheel at the contact with ground as follows:

$$F_{Vm} = k_2 F_{Vor} \tag{1}$$

Where: k_2-dynamic load factor for the static strength, shown in Figure 3; F_{Vor} -full of static load on the front wheels.

(2) The maximum braking power condition, the rear wheel brake load is:

$$F_{Br} = 0.8 F_{Vor} \tag{2}$$

$$F_{VBr} = F_{Vor} \tag{3}$$

Where: F_{Vor}-full static load for rear wheel; F_{Br}-one side braking force for the rear wheels; F_{VBr}-vertical force of rear wheels during braking condition.

(3) The maximum longitudinal force conditions, the rear wheel vertical load increases, the rear wheel load cases as follows:

$$F_{A2r} = 1.1 F_{Vor} \tag{4}$$

$$F_{VA2r} = k_1 \cdot F_{Vor} \tag{5}$$

Where: F_{A2r}-a longitudinal force on the rear wheel; k_1-vertical dynamic load factor for the durability, shown in Figure 3; F_{VA2r}-vertical load on the rear wheels during acceleration.

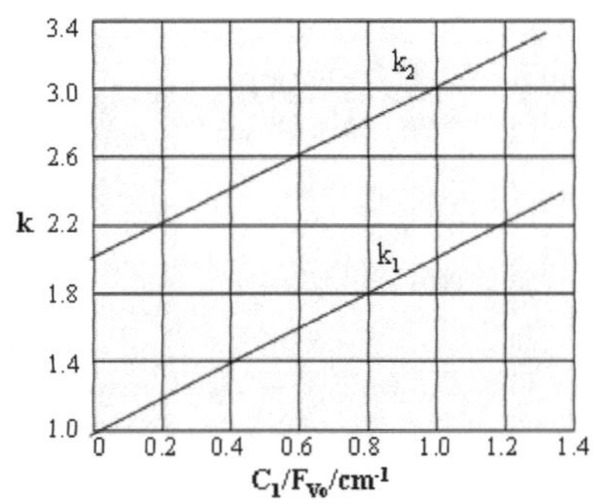

Figure 3. Characteristic curve of vertical dynamic load
(k_1- vertical dynamic load factor for the durability, k_2- dynamic load factor for the static strength)

(4) The maximum lateral force condition, the load on the rear wheels shown in Figure 4.

$$F_{Zrl} = \frac{m_2 g}{2} + \Delta F_Z \tag{6}$$

$$F_{Zrr} = \frac{m_2 g}{2} - \Delta F_Z \tag{7}$$

$$\Delta F_Z = \frac{m_2 a_y h}{t} \tag{8}$$

Where: m_2-the front axle load; ay-the turning acceleration; h-centroid height; t-wheelbase;
Therefore, lateral load F_{yrl}, F_{yrf} on the left and right side wheels are:

$$F_{yrl} = \mu F_{Zrl} \tag{9}$$

$$F_{yrr} = \mu F_{Zrr} \tag{10}$$

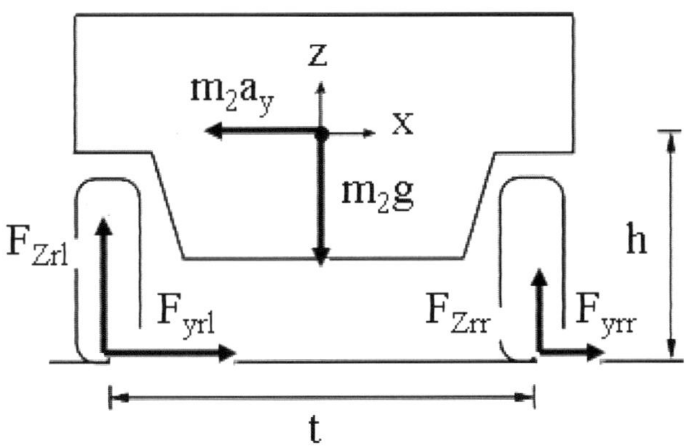

Figure 4. Force analysis diagram of automobile during cornering

3.Analysis of the overall strength of the Suspension System

The basic parameters of a car: wheelbase 2530mm; full centroid height 550mm; tread 1550mm; tire radius 289mm; rear-sided full load 381kg. Torsion trailing arm suspension system for the material parameters: rear beam material QSTE420TM; vertical arm material STEEL10; hub carrier scaffold is ST52-3; the shock absorber bracket, trailing arm and rear reinforcing plate reinforcing plate materials are SPHE. Where, the ultimate strength for material of QSTE420TM, STEEL10, ST52-3 and SPHE respectively is 471MPa, 424MPa, 355MPa and 300MPa.

Force exerted on wheels is rather complicated to analyze. In addition, material nonlinearity, geometric nonlinearity and contact nonlinear, together accounting for vast resources to be calculated, are all involved in finite element analysis, which can hardly be completed in a short term. Consequently, the role of the tire is not reckoned in this model. Instead, equivalent force taking place on a connection point between the wheel hub and the wheel is replaced. Constraints are: six degrees of freedom between vertical arm and body connection point are constrained; X, Y and Z directions of translational degrees of freedom between damper and body fixed point are constrained. Point load is applied in the hub of the wheel assembly point. According to the conditions considered the hub assembly center in the equivalent load conditions, as shown in Table 1:

Table1. Load at the center of the wheel rim assembly in typical operating conditions

Load case	F_X (N)	F_Y (N)	F_Z (N)	M_X(N • mm)	M_Y(N • mm)	M_Z(N • mm)
Load case 1	0	0	9335	0	0	0
Load case 2	2011	0	2011	0	-581179	0
Load case 3	3733	0	3733	0	-1078837	0
Load case 4	0	6622	6622	1913758	0	0

Under conditions (1), the impact load on the wheel shown in Figure 3 of vertical dynamic load curve, k_2 take 2.5. Vertical load on the rear wheels by the formula (1) be regarded as the 9335N, as shown in Table 1.Overall stress contours of rear beam shown in Figure 5, the maximum stress of 213MPa, appears in welding position between rear beam and vertical arm; Longitudinal arm stress contours, as shown in Figure 6, the maximum stress of 385MPa, appears in the bent cross section near the rear beam and trailing arm. Stress contours of absorber bracket as shown in Figure 7, the maximum stress of 274MPa, occur at the transition arc; Hub carrier bracket stress contours shown in Figure 8, the maximum stress of 227MPa, occur in small sectional place; Vertical arm reinforcing plate stress contours shown in Figure 9, the maximum stress 277MPa, appears in the longitudinal arm weld region; Rear beam reinforcing plate stress contours shown in Figure 10, the maximum stress of 225Mpa, appears in the curvature region. In this condition, the maximum stresses of the components do not exceed the yield strength of the materials to meet the static strength requirements.

Likewise, the maximum stress value and its distribution position of the components of suspension system in other three typical working conditions are acquired, using the same calculation method applied in condition 1.

By calculating under the four typical operating conditions, the conclusion is: in the maximum lateral force conditions, the maximum stress of the side connecting hub carriers and reinforcing plate of vertical arm is 409MPa, exceeding the yield limits. Under other conditions, the components of suspension systems meet the strength requirements.

4. Comparative Analysis

Results from the overall strength of the suspension system in finite analysis are compared and contrasted with previous records on single component. Figure 11 shows rear beam stress contours in a single piece of analysis. Table 2 shows the maximum stress of the two methods compared.

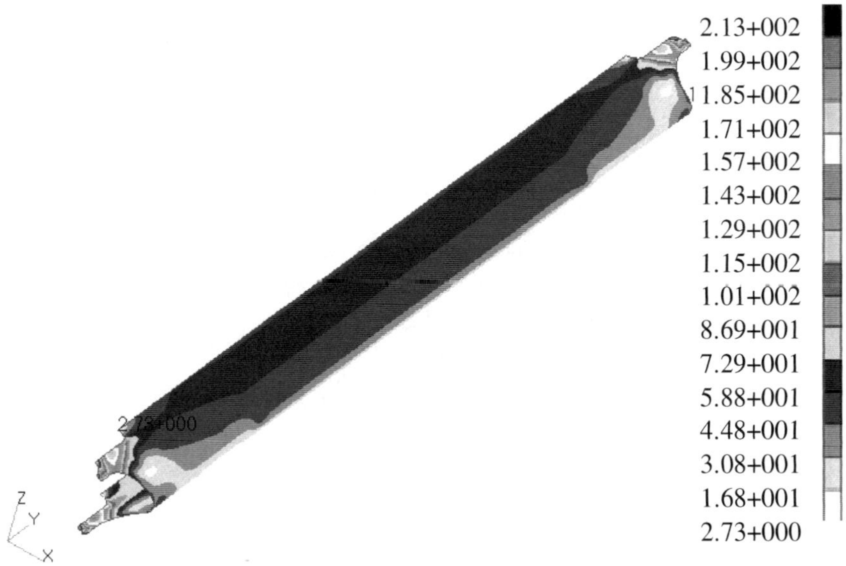

Figure 5. Stress contours of rear beam

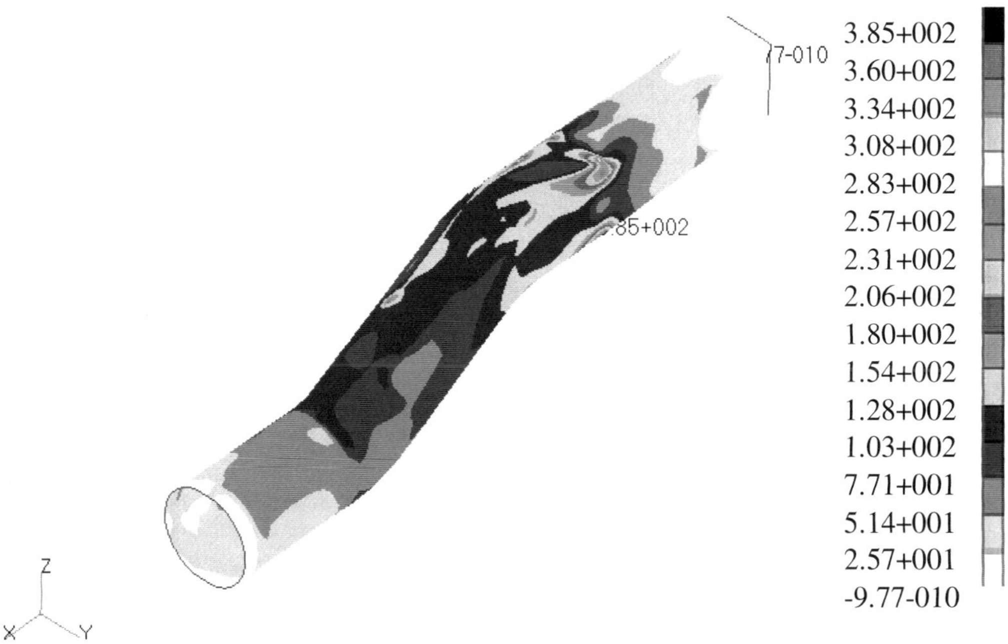

Figure 6. Stress contours of vertical arm

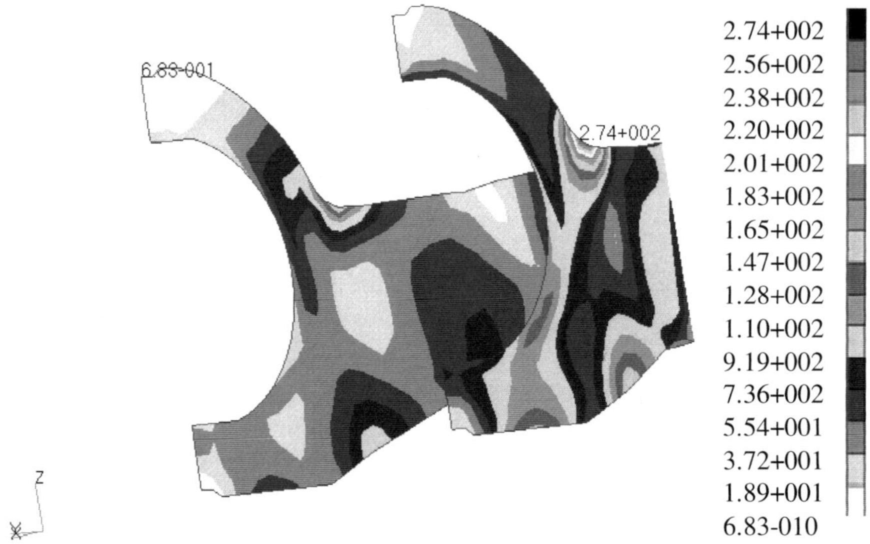

Figure 7. Stress contours of shock absorber bracket

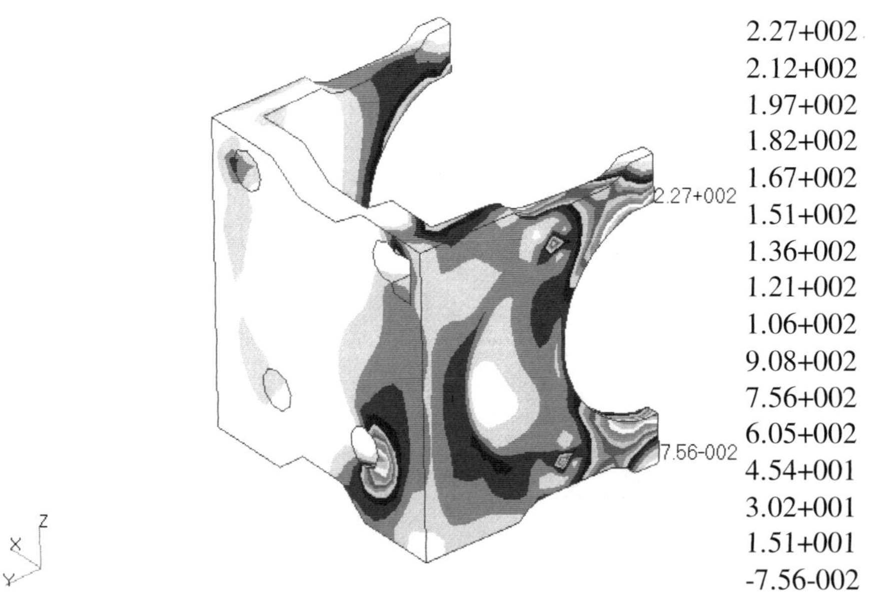

Figure 8. Stress contours of hub carrier bracket

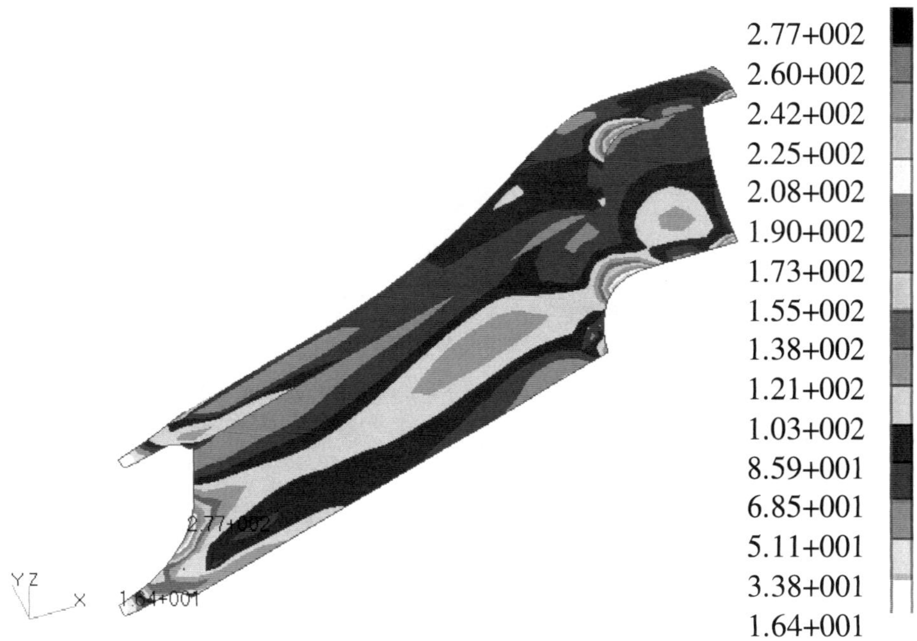

Figure 9. Stress contours of vertical arm reinforcing plate

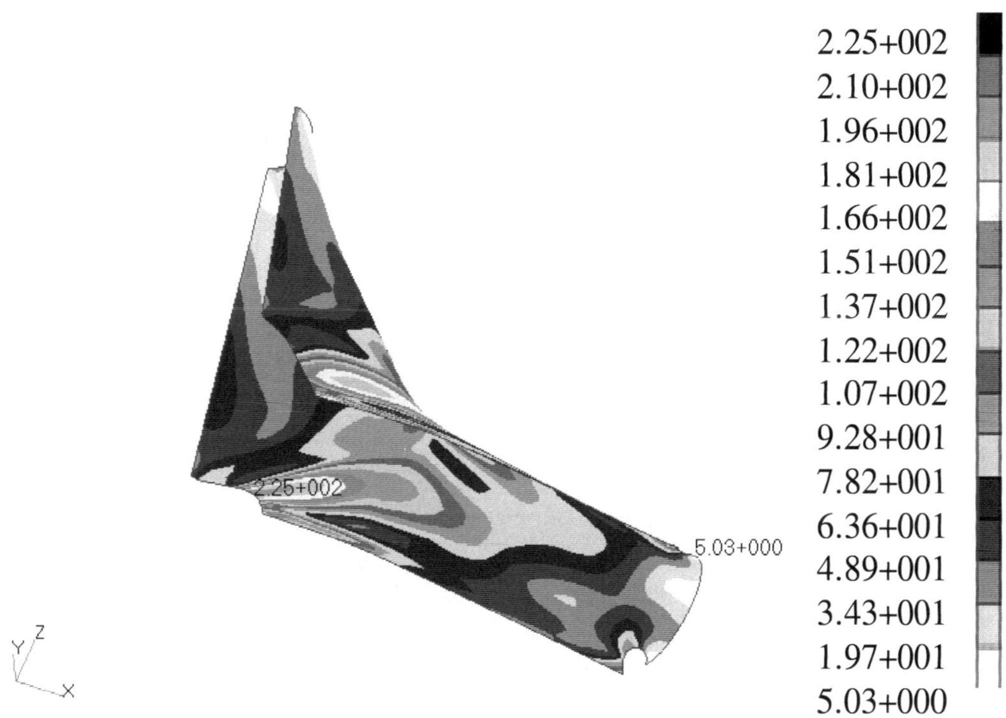

Figure 10. Stress contours of rear spar reinforcing plate

Figure 11. rear beam stress contours in a single piece of analysis

It is concluded from the charts that stress value on one-piece analysis is generally higher than the overall. The reasons are summarized below: (1) Connections between components are not considered in one-piece analysis, so differences are shown between simulation results and actual load. (2) Fixed constraint and concentrated load are applied in single-piece analysis, resulting in higher stress level. (3) The flexible rigidity of the sleeve connecting is not taken into account in the strength analysis of the suspension system, which has a certain degree of impact on the results.

Table 2. The maximum stress of the two methods compared

Type		Maximum stress results (MPa)			
		1	2	3	4
Rear beam	Single piece	265	77	141	386
	Overall	213	63	115	365
Vertical arm	Single piece	420	105	189	412
	Overall	385	86	161	375
Shock absorber bracket	Single piece	331	107	195	237
	Overall	274	98	169	218
Hub carrier bracket	Single piece	254	103	195	451
	Overall	227	90	166	409
Vertical arm reinforcing plate	Single piece	308	186	293	304
	Overall	277	161	279	273
Rear spar reinforcing plate	Single piece	269	97	164	317
	Overall	225	73	132	278

5.Conclusion

Stress of connecting bracket of the trailing arm hub seat suspension system exceeds the ultimate strength of the material. It is recommended to flatten depression while increasing the thickness of the plate. The rest of the components meet strength requirements under the four typical working conditions. It is concluded from comparative analyzing that stress values in single-piece analysis is generally higher than overall. The reasons include ignoring connections between components, the use of fixed constraint and application of concentrated load. Overall, the close-to-fact results from overall analysis, compared with rather conservative single-piece analysis, provide more accurate scientific proof for the design of suspension system.

6. Acknowledgments

The work presented in this paper has been supported by：Research Program of Students Maker Laboratory Construction Plan of Anhui Provincial Quality Project No. 2015ckjh144.and Research and design of electric vehicle braking energy recovery technology and braking energy recovery test bed No.2016zdzr08.and the characteristic speciality of Automobile technical service and marketing No. 20 15tszy081and Research on the integrated control method of vehicle chassis system based on multi-body dynamics model of 2017 anhui university science key research project No.KJ2017A753

7. References

[1] Xiaofeng Wang. Chassis Design of Automobile [M]. Beijing: Tsinghua University Press, 2010, 47-53, 272-273, 282-294
[2] Donghua Chen,Xiaoxiong Jin. Dynamics Simulation and Experimental Study on Torsion Bar rear Suspension of Passenger Car [J]. Automotive Engineering, No.2, Vol 29, 2007
[3] Yanhua Liu. Development and Research of Twist Beam Rear Suspension [J]. JOURNAL OF SHENYANG UNIVERSITY, No.4, Vol.18, 2006
[4] Mingjun Luo, Zhichao Hou. Simulation of Load Time History on Spindle Nose and Fatigue Life Assessment of Rear Torsion Beam [J]. Computer Simulation, 2014, 31(1): 180-184
[5] Yiping Shi, Yurong Zhou. ABAQUS Finite Element Analysis Example Explanation [M]. Mechanical Industry Press, 2013, 226-276

Vibration Tracking Control of Piezoelectric Bimorph Bender in Novel Reciprocal State Space Form

Yuan-Wei Tseng[1], **Rong-Ching Wu** [1] **(correspondence) and Chia-Chuan Tsai**[1]

[1] Dept. of Electrical Engineering, I-Shou University, Kaohsiung City, Taiwan R.O.C.

E-mail: yuanwei@isu.edu.tw, rcwu@isu.edu.tw (correspondence), cdefcdef0102@gmail.com

Abstract. In this paper, the deflection of a piezoelectric bimorph bender is controlled to track given time varying reference command solely utilizing direct acceleration measurement without integration. Novel estimator that directly applies acceleration measurement and controller that applies estimated state derivative signals have been developed in reciprocal state space (RSS) form. Simulations for the augmented system of both closed loop system and estimator have been carried out to successfully verify the proposed methods. The design approach in this paper is applicable to smart structures with accelerometers as sensors.

1. Introduction

Piezoelectric bimorph bender is a smart structure and can serve as both sensor and actuator in many electromechanical applications such as acoustic sensor, force sensor, vibration sensor [1], energy harvesting [2-3], minute robotic [4], and actuators [5]. Typically, the structure of a piezoelectric bimorph bender [6-7] is an elastic plate sandwiched between two piezoelectric plates and is mounted in a cantilever arrangement. In this paper, an accelerometer is installed on a piezoelectric bimorph bender as sensor. A novel estimator is designed based on the direct acceleration measurement. In addition, a novel tracking controller based on estimated state derivative is designed so that the deflection of the bimorph piezoelectric plate can track the time varying reference command. However, the proposed design objectives are not achievable applying traditional state feedback algorithms in the state space form as follows.

$$\dot{x} = \overline{A}x + \overline{B}u \quad (1a)$$

$$u = -kx \quad (1b)$$

$$y = Cx \quad (1c)$$

, where x, u and y are state, control and measurement, respectively. The reason is that acceleration measurement only can be modelled as state derivative. Therefore, people cannot apply available state related measurement feedback algorithms without using numerical integrations which will increase complexity and implement cost of controller.

To directly utilize state derivative measurement in controller and estimator designs, the first author proposed the following reciprocal state space (RSS) form [8].

$$x = A\dot{x} + Bu \quad (2a)$$

$$u = -k\dot{x} \quad (2c)$$

$$y = C\dot{x} \quad (2c)$$

The name of RSS form was given because the open loop poles of system (2a) are the reciprocals of the eigenvalues of matrix A. In this paper, the vibrational tracking control of a piezoelectric bimorph plate is carried out in RSS form applying state derivative measurement in design.

2. Modelling of a smart piezoelectric bimorph bender

The smart piezoelectric bimorph bender in this research consists of a composite plate with piezoelectric actuation and a micro-machined accelerometer as sensor. We developed discrete model of the smart plate for both dynamic analysis and control designs of transverse vibration. The candidate composite plate in [9] has laminated design angles of [-55.66/43.62/4.18] and has six graphite/epoxy piles with length, width and thickness of 0.305, 0.076 and 0.134×10^{-2} m, respectively. We model the smart plate in such a way that, 0.4 mm piezoelectric PVDF (polyvinylidene fluoride) material is distributed on both upper and lower surfaces of composite plate and one electrode serving as actuator is installed on upper surface while one micro-accelerometer serving as sensor is attached on the lower surface. Applying Hamilton's law and using the first four shape functions in [10] and the physical properties of PVDF in [11], three sub-models are first built for the structural, electrical and coupling subsystems and then assembled to get the resultant discrete model as follows.

$$\begin{bmatrix} M_s+M_{pz} & 0 \\ 0 & 0 \end{bmatrix}\begin{bmatrix} \ddot{u} \\ \ddot{u}_e \end{bmatrix}+\begin{bmatrix} K_s+K_{pz} & -K_{u\phi} \\ K_{\phi u} & K_{\phi\phi} \end{bmatrix}\begin{bmatrix} u \\ u_e \end{bmatrix}=\begin{bmatrix} 0 \\ \Gamma q \end{bmatrix} \tag{3a}$$

$$y=C_a\ddot{u} \tag{3b}$$

, where M_s and K_s are mass and stiffness matrices of the composite plate while M_{pz}, K_{pz}, $K_{u\phi}$ and $K_{\phi\phi}$ are PVDF mass and PVDF stiffness, coupling and capacitance matrices, respectively.

In addition, Γ is related to the location of electrode and electric shape function which consistent with prescribed voltage/charge boundary conditions. Furthermore, u, u_e and Γq are the generalized displacement vector, voltage across the PVDF and surface charge, respectively. Based on the location of micro-accelerometer and the first four mechanical shape functions in [10], one can get the sensor matrix C_a and acceleration measurement y. Details of modelling methods and techniques for smart plate are available in [12]. From the second equation in (3a), one can have

$$u_e=K_{\phi\phi}^{-1}\Gamma q-K_{\phi\phi}^{-1}K_{\phi u}u \tag{4}$$

Substituting (4) into first equation in (3a), we have the following system in second order form.

$$(M_s+M_{pz})\ddot{u}+(K_s+K_{pz}+K_{u\phi}K_{\phi\phi}^{-1}K_{\phi u})u-K_{u\phi}K_{\phi\phi}^{-1}\Gamma q=M\ddot{u}+Ku+Df=0 \tag{5}$$

, where $D=-K_{u\phi}K_{\phi\phi}^{-1}\Gamma$ and $f=q$.

3. Estimator design in RSS form

Defining $x^T=[u\ \ \dot{u}]^T$, system equation (5) is expressed in reciprocal state space form as follows.

$$x=\begin{bmatrix} 0 & -K^{-1}M \\ I & 0 \end{bmatrix}\dot{x}+\begin{bmatrix} -K^{-1}D \\ 0 \end{bmatrix}f=A\dot{x}+Bf \tag{6a}$$

$$y=[0\ \ C_a]\dot{x}=C\dot{x} \tag{6b}$$

To directly utilize the sensed acceleration signal without numerical integration, the following estimator in RSS form is proposed.

$$\hat{x}=A\dot{\hat{x}}+Bf+L(y-\hat{y}) \tag{7a}$$

$$\hat{y}=C\dot{\hat{x}} \tag{7b}$$

, where $\hat{x}$ and $\dot{\hat{x}}$ are the estimated state and estimated state derivative, respectively.

The estimated state error is then defined as:

$$e=\hat{x}-x \tag{8}$$

Subtracting (6a) from (7a), we get the following error dynamics of estimator in RSS form.

$$e = (A - LC)\dot{e} \quad (9)$$

Therefore, once the estimator gain L in (9) is properly designed, the estimation error will converge to zero. Consequently, full estimated state derivative feedback gain k in (10) is designed so that the closed loop system of smart piezoelectric bimorph bender is stable, namely, all system eigenvalues have negative real parts.

$$f = -k\dot{\hat{x}} \quad (10)$$

Combining (6)-(10), an augmented system equation of state and estimator error is obtained as follows.

$$x_t = \begin{bmatrix} x \\ e \end{bmatrix} = \begin{bmatrix} A - Bk & -Bk \\ 0 & A - LC \end{bmatrix} \begin{bmatrix} \dot{x} \\ \dot{e} \end{bmatrix} = A_t \dot{x}_t \quad (11)$$

Since A_t in (26) is an upper triangle matrix, full estimated state derivative feedback gain k and estimator gain L can be designed separately. The separation principle also holds for systems in reciprocal state space form. For the reciprocal state space system in (11), the eigenvalues of $A - Bk$ should be designed as the reciprocals of desired closed loop eigenvalues while the eigenvalues of $A - LC$ should be designed as the reciprocals of desired estimator eigenvalues.

4. Tracking control design in reciprocal state space form

The following tracking controller in (14) can work for the RSS system in (12-13) in following discussion.

Plant: $x = A\dot{x} + Bf$ (12)

Performance output: $z = Hx$ (13)

Tracking controller: $f = -k\dot{\hat{x}} + Nr$ (14)

, where x, $\hat{x}$, y, u and r are state vector, estimated state vector, sensor measurement vector, control vector and reference command vector, respectively while z is the system performance output of interest, which must track the given reference command r. In (14), feedback gain k is first design to build a stable closed loop system, then feedforward gain N is designed to track the reference command. Note that if feedback gain k is not properly desgn and the closed loop system is not stable, no matter what feedforward gain N is designed, the performance output z can never well track the reference command r.

In this paper, pole placement method in RSS form [13] is applied to design feedback gain k to ensure that the closed loop system is stable. When we applies pole placement design for RSS system in (12), the gain k must be designed such that the eigenvalues of $(A - Bk)$ are reciprocals of the desired closed loop eigenvalues. In addition, LQR [13] and sliding mode control [14] in RSS form developed by the first author can also be used to build a stable closed loop system.

When the closed loop system is stable ($\dot{\bar{x}}(\infty) = 0$ and $\dot{\bar{e}}(\infty) = 0$) we have steady state as

$$\bar{x} = (A - Bk)\dot{\bar{x}} + BK\dot{\bar{e}} + BNr_0 = BNr_0 \quad (15)$$

When the steady state of tracking error is zero, we have

$$\bar{\varepsilon}(\infty) = r_0 - H\bar{x} = r_0 - HBNr_0 = (I - HBN)r_0 = 0 \quad (16)$$

Consequently, we have feedforward gain N as

$$N = (HB)_{right}^{-1} \quad (17)$$

, where $(HB)_{right}^{-1}$ is the right inverse of matrix HB.

Suppose that HB is a $m \times n$ full rank matrix and $m \leq n$. We have

$$(HB)_{right}^{-1} = (HB)^T (HB(HB)^T)^{-1} \quad (18)$$

Since feedforward gain N in (17) is independent of feedback gain k, the feedback gain k and feedforward gain N can be designed separately.

5. Numerical Example

The matrices of the model of piezoelectric bimorph bender in (5) are modelled as follows.

$$M = 10^{-3} \times \begin{bmatrix} 287 & 4 & 0 & 0 \\ 4 & 367 & 0 & 0 \\ 0 & 0 & 24.72 & 0.189 \\ 0 & 0 & 0.189 & 23.267 \end{bmatrix}, K = \begin{bmatrix} 38.1 & -0.2 & -13.4 & -16.4 \\ -0.2 & 1510.3 & 96.9 & -167.1 \\ -13.4 & 96.9 & 659.0 & -849.6 \\ -16.4 & -167.1 & -849.6 & 5250.9 \end{bmatrix},$$

$D^T = [348.1431 \quad 424.7985 \quad 0 \quad 0]$, $C_a = [1.315 \quad 0.27 \quad 0.631 \quad 0.081]$

Suppose that the first generalized displacement is of interest, the performance output matrix H is then selected as follows.

$H = [1 \quad 0 \quad 0 \quad 0 \quad 0 \quad 0 \quad 0 \quad 0]$

Substituting the above matrices to the reciprocal state space system in (6a). One can verify that the open loop system has zero damping and cannot track step command and time varying command $2\sin 5t$ by applying feedforward gain N alone as shown in figure 1 and 2.

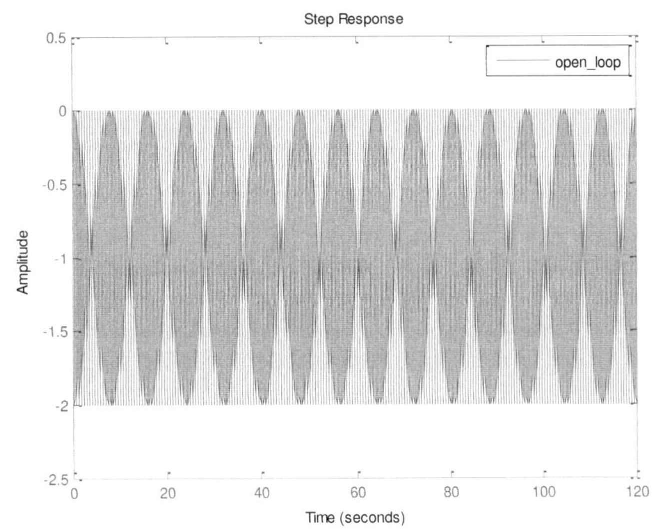

Figure 1. Open loop step tracking

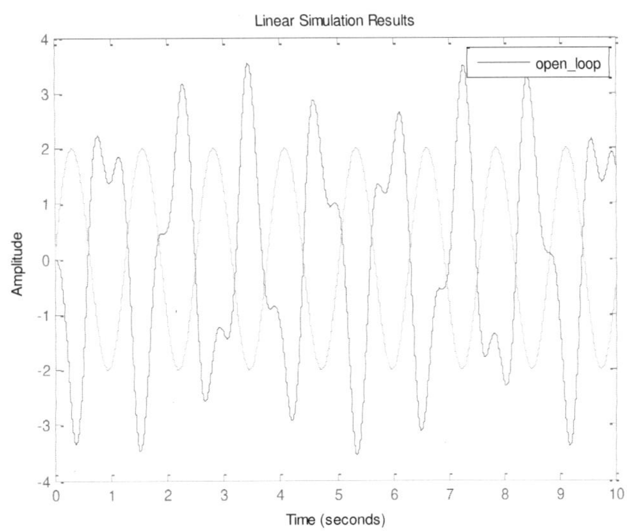

Figure 2. Open loop $2\sin 5t$ tracking

The desired poles of closed loop system are selected as $-1\pm 2i, -2\pm 2i, -3\pm 4i$, $-4\pm 6i$ for both estimator and controller. Therefore, feedback gain k in (14) and estimator gain L in (7a) are designed through pole placement method in such a way that the eigenvalues of $(A-Bk)$ and $(A-LC)$ are the reciprocals of those desired poles. Consequently, the augmented system in (11) is built for simulation.

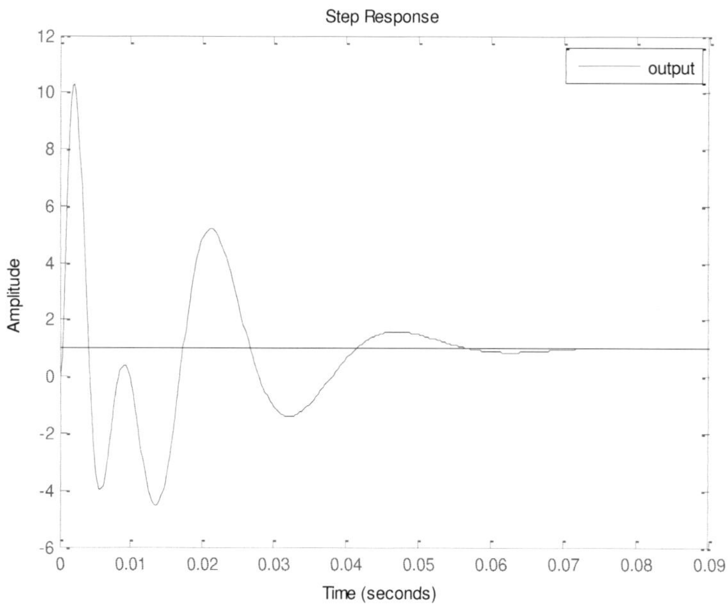

Figure 3. Closed loop step tracking

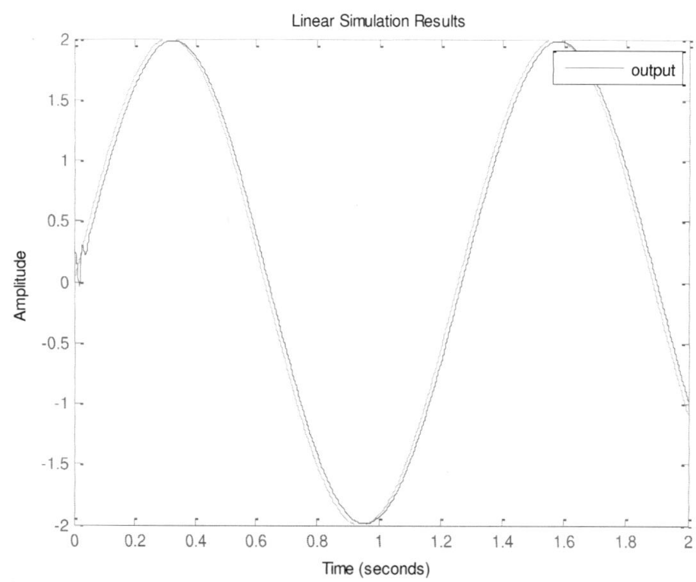

Figure 4. Closed loop $2\sin 5t$ tracking

As anticipated, the proposed design can well track step command and time varying command $2\sin 5t$ as shown in figure 3 and 4. Therefore, the proposed design approach is successfully verified.

6. Conclusion

In this paper, the designs of novel estimator that directly employs acceleration measurement and tracking controller that applying estimated state derivative feedback in RSS form have been introduced. Simulations based on a model of piezoelectric bimorph bender have been performed to verify the proposed design approach. The piezoelectric bimorph bender that can deflect itself by following time varying reference command can be used as a lifting surface and communication antenna of aerospace vehicles and in many other applications. The proposed design approach is also suitable for applications of other smart structure with accelerometers as sensors.

7. Acknowledgments

Authors would like to thank the Ministry of Science and Technology, Taiwan, ROC for the financial support of this research under grant MOST 106-2221-E-214-008.

8. References

[1] Jacob Fraden, 2003 *Handbook of Modern Sensors: Physics Design and Applications* (New York: AIP Press)

[2] Zheng X etc., 2015 Analysis of Energy Harvesting Performance for d15 Mode Piezoelectric Bimorph in Series Connection Based on Timoshenko Beam Model, *IEEE/ASME TRANSACTIONS ON MECHATRONICS*, **20**, 728-739.

[3] Zhang H, Afzalul K, 2014 Design and Analysis of a Connected Broadband Multi-Piezoelectric-Bimorph-Beam Energy Harvester, *IEEE Transactions on Ultrasonics, Ferroelectrics, and Frequency Control*, **61**, 1016-1023.

[4] Cen L and Erturk1 A, 2013 Bio-inspired Aquatic Robotics by Untethered Piezohydroelastic Actuation, *Bioinspiration & Biomimetics*, **8**, 016006

[5] Inoue J-I, Kanda K, Fujita T and Maenaka K, 2015 Thin-film Piezoelectric Bimorph Actuators with Increased Thickness using Double Pb[Zr,Ti]O3 Layers, *Journal of Micromechanics and Microengineering*, **25**, 055001.

[6] Rios S and Fleming A, 2016 Design of a Charge Drive for Reducing Hysteresis in a Piezoelectric Bimorph Actuator, *IEEE/ASME TRANSACTIONS ON MECHATRONICS*, **21**, 51-54.

[7] Ajitsaria J, Choe S, Shen D and Kim D, 2007 Modelling and Analysis of a Bimorph Piezoelectric Cantilever Beam for Voltage Generation, *Smart Mater. Struct.*, **16**, 447–454.

[8] Tseng Yuan-Wei, 1997 *Control Design of Linear Dynamic System with Matrix Differential Equations for Aerospace Applications*, Ph. D. Dissertation, in Department of Aerospace Engineering, The Ohio State University: Columbus, USA.

[9] Dracopoulos T and Oz H, 1992 Integrated aeroelastic control optimization of laminated composite lifting surfaces, *Journal of Aircraft*, **29**, 280-288.

[10] Hollowell S and Dugundji J, 1984 Aeroelastic flutter and divergence of stiffness coupled, graphite/expoxy cantilevered plates, *Journal of Aircraft*, **21**, 69-76.

[11] Diwekar A M and Yedavalli R K, 1996 Smart structure control in second order form, *Journal of Smart Structures and Materials*, **5**, 429-436.

[12] Yedavalli R K, 1998 New Control Design Techniques Tailored to Smart Structure System, *Air Force Research Laboratory Report AFRL-VA-WP-TR-1998-3080*, Wright-Patterson Air Force Base, Ohio, USA.

[13] Tseng Yuan-Wei, 2008 Control Designs of Singular Systems Expressed in Reciprocal State Space Framework with State Derivative Feedback, *International Journal of Control Theory and Applications*, **1**, 55-67.

[14] Tseng Yuan-Wei, Wang Yu-Ning, 2013 Sliding Mode Control with State Derivative Output Feedback in Reciprocal State Space Form, *Abstract and Applied Analysis*, 1-12. https://www.hindawi.com/journals/aaa/2013/590524/

Silicone Grease for Outdoor Gas-insulated Switchgear (GIS) Application

Liang Song[1, a], Feng Ying[2, b], Hui Zhang[3, c], and YiXi Liu[4, d]

[1,3]Corporate Research Center, ABB (China) Limited,
BEZIT Park Block 103 Level 1, Jiuxianqiaobei Road, no.10, Chaoyang District, Beijing, P.R. China
[2,4]ABB High Voltage Switchgear (Xiamen) Co., Ltd,
Guanghui Building, 6# Chuang Xin 3rd Road, Xiamen S.E.Z Fujian, P.R. China

E-mail: [a]William-liang.song@cn.abb.com and [c]Feng.Ying@cn.abb.com

Abstract. Silicone grease are widely applied in flange connections of gas-insulated switchgear (GIS) providing protection against flange and sealing O-ring corrosion and subsequent SF6 gas leakage. During GIS operation, grease bleeding phenomenon was frequently observed, which not only result in a poor appearance of flange surface but a potential SF6 leakage in the long-term application. In this paper, the possible reasons for this phenomenon is briefly discussed. In addition, the key parameters and outdoor performance of several commercial grease products were verified for a better understanding of its temperature resistance. It is shown that some grease products can have a better control on bleeding performance and should be more suitable for the outdoor GIS application.

1. Introduction

Flange corrosion in outdoor gas-insulated switchgear (GIS) often leads to SF6 gas leakage which in severe cases could necessitate a shutdown. To avoid the corrosion and eliminate the gas leakage, flange connections are normally treated in the factory or during installation on site by injection of silicon grease into the flange connection via an injection nipple provided in the flange [1]. With this treatment the flange connections can be protected against corrosion that caused when in contact with the air or moisture, as illustrated in Figure 1[2]. In the meantime, silicone grease are also applied on assembled O-rings in order to protect it from damage by abrasion. It helps seat the O-ring properly, speeds up assembly operations, and makes automated assembly line procedures possible.

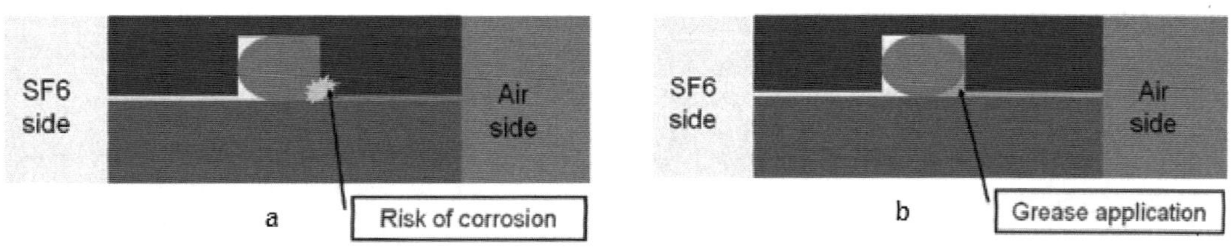

Figure 1. a. Risk of corrosion on sealing; b. Risk mitigating using silicone grease

Generally, silicone grease is a solid to semifluid product that consists of a dispersion of thickening agent in a liquid lubricant [3]. In an application, a silicone grease gradually releases oil into the working areas of the machine surfaces in order to lubricate them. This oil release phenomenon is normally named 'oil bleed', one possible reason of which is related to the grease formulation. As the base oil content is increased and the amount of thickener system is decreased, the forces of attraction also decrease, thus resulting in the base oil being loosely held in the thickener system matrix and easily separated. Therefore, it is important for a grease to have the proper balance of base oil and thickener system content to function properly. Another possible reason may relate with the storage or even the transport conditions, such as high temperature, high moisture, long duration and vibration etc. The severe outdoor environment can probably speed up the bleeding phenomenon and extent. In this paper, several key parameters of some commercial grease products were firstly compared to screen candidate grease products with proper formulation. A subsequent environmental tests such as neutral salt spray, high-low temperature cycle and long-term outdoor test were then conducted with selected grease to verify their applicability in outdoor GIS.

2. Experimental

2.1. Material

The commercial silicone grease and O-rings (NBR and EPDM) were provided by varied suppliers, separately.

2.2. Neutral Salt Spray Test

The neutral salt spray (NSS) test was performed with a duration of 720h at 35°C±2°C and a pH of (6.5~7.2) in accordance with ISO 9227 [4]. After NSS test, the flange modules were disassembled accordingly and the grease samples were further characterized for possible structure change.

2.3. High-low Temperature Cycle

The high-low temperature cycle test was performed with GIS modules filled with SF6 gas [5]. The temperature range is selected as -40°C to 120°C in accordance with GB/T 11022 [6]. The gas tightness was tested before and during the temperature cycles. The module surface and groove were then checked if any grease bleeding or missing after completing the thermal cycle test.

2.4. Outdoor Aging Test

The GIS products assembled with selected grease were placed in actual outdoor conditions for more than 6 months and the flange surfaces were visually checked in a certain time interval as well as the gas tightness.

3. Results and Discussion

As mentioned previously, a grease releases its base oils from the thickener system when it is squeezed or stressed, just like a sponge that release water when it is squeezed. The stresses a grease encounters can be generated thermally during application or storage. Therefore, the thermal resistance is of importance in selecting proper grease. Here, some key parameters of several silicone grease samples were tested and summarized in Table 1. It is to be noted that oil separation represents colloidal stability, which is the resistant ability of separating oil in the long-term storage and use. The less the oil separation value, the less possibility of bleeding phenomenon. From Table 1, it can be seen that Grease 4 and Grease 5 show a better comprehensive performance as compare to other grease samples. Therefore, Grease 4 and Grease 5 were chosen for the following tests.

Table 1. Key parameters of silicone grease

Test item	Dropping point [℃]	Oil separation, steel mesh, (120℃×30h) [%]	Evaporation loss (120℃×22h) [%]
Test method	GB/T 4929	SH/T 0324	GB/T 7325
Grease 1	205	4.4	0.18

Grease 2	>260	0	0.63
Grease 3	187	0.18	0.16
Grease 4	>260	0	0.36
Grease 5	>260	0	0.37

The aluminum flanges were prepared and further assembled with different combinations of grease and O-rings. The size of the flange is about 215mm in diameter. The assembled aluminum flanges were carefully checked and marked then put into salt spray chamber, as shown in Figure 2.

Figure 2. Layout of flange samples in salt spray chamber

A visual inspection was carried out on the surface of aluminum flanges at a certain time. Figure 3 illustrated an example of flange module assembled with Grease 5 and NBR O-ring after NSS test for 720h. As can be seen in Figure 3, the silicone grease were leaked out from the screw thread and partially from the interface of the flange. Similar phenomenon can also be found in other assembled flanges.

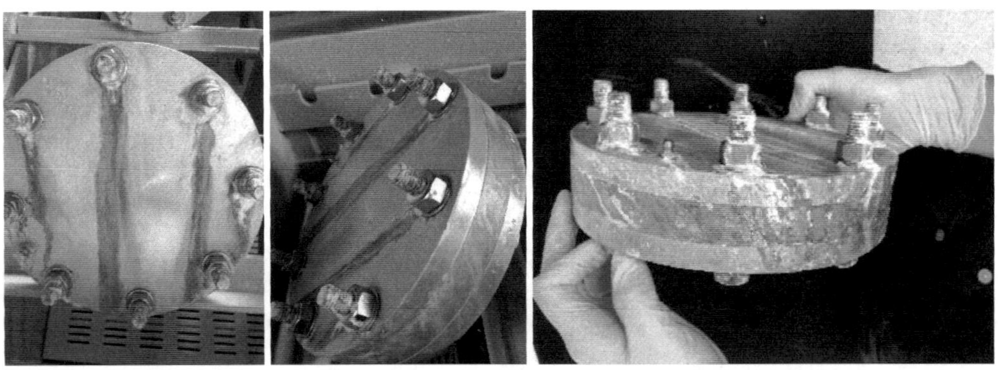

Figure 3. Assembeld flange with Grease 5 and NBR O-ring after NSS test for 720h

All flanges were taken out of the chamber and then disassembled to check the inner surface. It turns out that no clear stain or contamination was found on inner surface but only partial of silicone grease were left over as a result of the leakage, as displayed in Figure 4. To verify the potential influence of NSS on the grease, the chemical structure were further analyzed by FTIR [7].

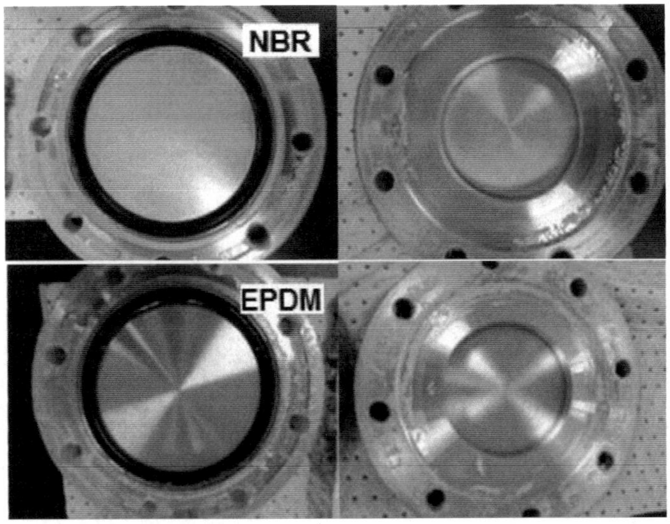

Figure 4. Inspection on dissembled flanges (Grease 5 and NBR/EPDM O-ring)

Figure 5 displays the FTIR spectrums of Grease 4 and Grease 5 before and after 720h NSS test. It can be seen that there is no obvious change in main characteristic peak position while only a very small change in peak intensity, indicating no structure alter as a result of the NSS test.

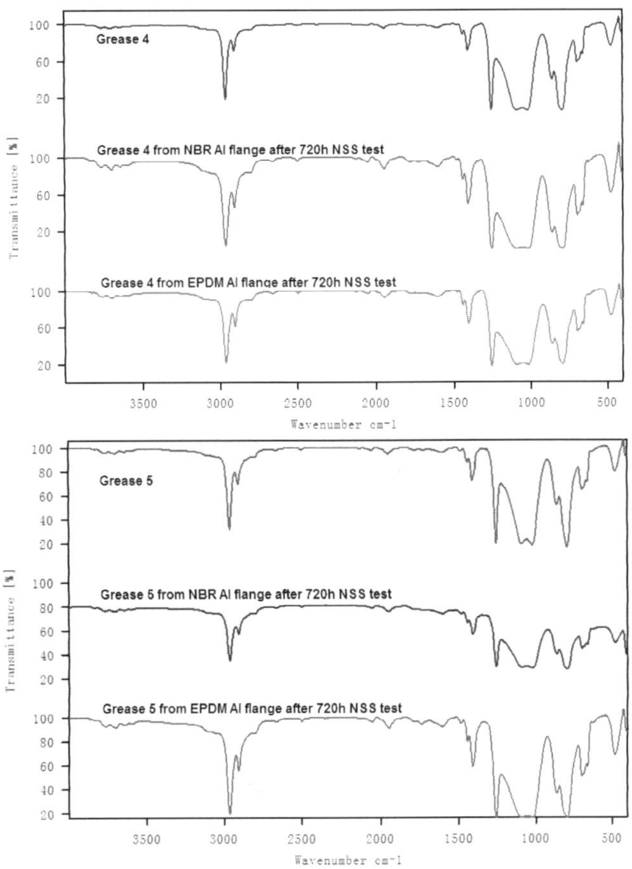

Figure 5. FTIR spectrum of Grease 4 and Grease 5 before and after NSS test

The high-low temperature cycle test was performed with GIS modules to further check the temperature resistance performance of grease. The temperature range is selected from -40°C to 120°C according to GB11022-2011. Figure 6a shows the drawing of the module with a total size of approximate 400 mm×400 mm. Before test, both flange ends were assembled with cover plate to form a closed shell, as displayed in Figure 6b. After thermal cycle test, the module was disassembled to check the grease status. Figure 5c shows an example of GIS module with Grease 4, from which it can be seen that the grease was distributed evenly with little grease loss in the groove.

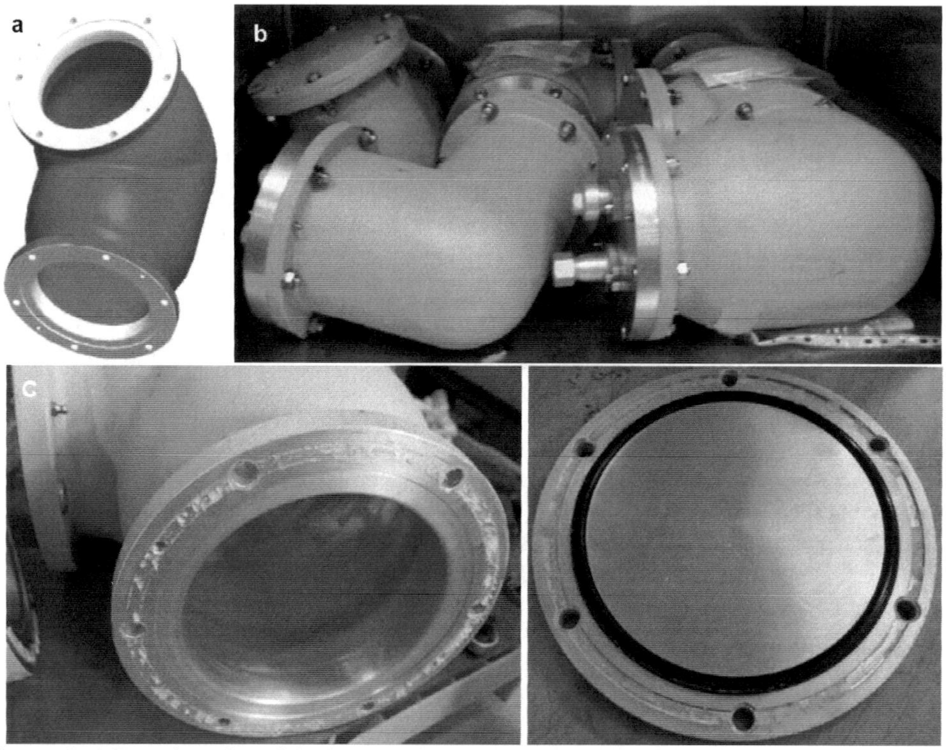

Figure 6. a) Drawing of module; b) Actual modules for the thermal cycle test; c) Example of disassembled flange to check the status of Grease 4

To better understand the long-term stability of the grease, an actual outdoor environment test was also conducted with GIS busduct, as shown in Figure 7. Two kinds of grease were injected into the flange groove by manual glue guns, respectively. The modules were then filled with SF6 gas. All GIS modules were placed outside for more than 8 months. For every 2 weeks, the status of module surface was recorded and the gas tightness was tested to see if any leakage happened. Results showed that the gas pressure was very stable and almost no loss after the test. Meanwhile, the disassembled module illustratrated that the grease were well kept in their original state without any loss or contaminant.

Figure 7. Outdoor verification test for GIS busduct

4. Summary

Silicone grease has been applied in outdoor GIS to avoid flange and sealing O-ring corrosion and lower the possibility of SF6 leakage. One problem during the application of silicone grease is oil bleeding, which may lead to serious loss of grease under severe environment and subsequently give rise to flange corrosion. With accumulated experience of GIS solutions, ABB has been in position to perform a detailed analysis on aging of grease products and to select the optimum grease materials to guarantee the long-term tightness of outdoor GIS application.

5. References

[1] G. Yang, C. Li, W. Jin and C. Zhang. High Voltage Apparatus, Vol. 49 (2013), p. 120
[2] A. Ficheux, D. Depres, E. Laruelle and Y. Kieffel. CIGRE (2012), C3-209
[3] L.G. Ludwig: *Proper storage and use of grease to avoid excessive oil bleeding and separation* (information on http://www.schaefferoilreps.com)
[4] ISO 9227 (2012), Corrosion tests in artificial atmospheres-Salt spray tests
[5] Test program for "High-low temperature cycle test of waterproof silicone grease", ABB HVS (Xiamen), 2014
[6] GB/T 11022-2011, Common specifications for high-voltage switchgear and controlgear standards
[7] E. Siwiec and E. Pawelec. CHEMIK, Vol. 65 (7), p. 633 (2011)

AUTHOR INDEX